A Laboratory Course in BIOMATERIALS

A Laboratory Course in BIOMATERIALS

Wujing Xian

CRC Press is an imprint of the
Taylor & Francis Group, an **informa** business

CRC Press
Taylor & Francis Group
6000 Broken Sound Parkway NW, Suite 300
Boca Raton, FL 33487-2742

Printed in the United States of America on acid-free paper
10 9 8 7 6 5 4 3 2 1

International Standard Book Number: 978-1-4200-7582-3 (Paperback)

Library of Congress Cataloging-in-Publication Data

Xian, Wujing.
A laboratory course in biomaterials / Wujing Xian.
p. ; cm.
Includes bibliographical references and index.
ISBN 978-1-4200-7582-3 (pbk. : alk. paper)
1. Biomedical materials--Laboratory manuals. 2. Biomedical materials--Study and teaching. I. Title.
[DNLM: 1. Biomedical Engineering--Laboratory Manuals. QT 25 X64L 2009]

R857.M3X53 2009
610.28'4--dc22 2009011787

Visit the Taylor & Francis Web site at
http://www.taylorandfrancis.com

and the CRC Press Web site at
http://www.crcpress.com

Contents

Preface

The area of biomedical engineering has vastly expanded in the past couple of decades. The need to educate the multidisciplinary students who will make up the work force in research and industry in biomedical engineering has also correspondingly increased. A brief survey of bioengineering and biomaterials programs in a number of academic institutions reveals a general consensus on topics that are covered by lecture courses. Topics and other teaching aids for laboratory courses, however, are much less readily available. In fact, few biomaterials lab courses exist at present, despite the clear need. It is the intention of this book to provide a laboratory curriculum that is comprehensive in scope as well as current in its perspective. This book is suitable for an undergraduate laboratory course in biomaterials, and bioengineering at the senior or junior level. It is also designed to help lower the barriers for entry into biomaterials for the more "traditional" engineering departments (such as materials science, chemical engineering, and mechanical engineering) since the cost in time and resources required for developing such laboratory courses can be quite high. This course is inherently multidisciplinary, integrates a variety of principles from materials science, mechanical engineering, chemistry, biochemistry, molecular and cell biology, and tissue engineering, and will train students in laboratory skills, data analysis, problem solving, and scientific writing. Experiments in this course are described in the form of flexible modules that can be chosen and adapted for the needs of different departments. Within each module, a range of multidisciplinary knowledge and laboratory practices are organized around a central theme, so that students can see the labs not as a compilation of procedures but rather as a coherent whole consisting of interconnections from various disciplines. Much has happened recently in this dynamic field; many experiments in this lab course are adapted from research papers and reflect recent progress in bioengineering and biomaterials.

As a laboratory manual, this book provides step-by-step descriptions of lab procedures, reagents, equipment, and even data processing guidelines so that a laboratory course can be started from scratch. These descriptions are guidelines rather than rigid prescriptions, and the experiments can be adapted according to the instructional laboratory settings and the students' learning needs. The questions following each module incorporate some of the frequently encountered problems and mistakes made by the students from my own teaching experience. Finally, it is my hope that this laboratory course is a fun and rewarding experience for students as well as teachers. Let's get started!

Acknowledgments

This book is based on the biomaterials laboratory course that I have taught for several years at the Materials Science and Engineering Department (MatSE) at the University of Illinois at Urbana-Champaign. I am very grateful to the people who have helped me with the course and with writing the book. Thanks are in order to Professor J. J. Cheng for the many helpful discussions; to Dr. Joanne Manaster, Sheeny Lan, Dr. Aylin Sendemir-Urkmez, and Dr. Sharon Wong for their help with cell cultures; to Dr. Raju Perecherla and Professor David Cahill for access to and help with instruments; to Spencer Shultz for his expert machine-shop work; and to Zach Culumber, Lanfang Li, Scott Slimmer, and Lihua Yang for their assistance. I am also thankful for the commitment to and support for biomaterials by my department. Especially I'd like to thank Jay Menacher, assistant to the head of MatSE, who has made things easy. And lastly, I am deeply grateful to Gerard Wong, who suggested that I write this book and helped with his expertise and moral support along the way: This book would not have been possible without him.

About the Author

Wujing Xian received a BS degree in chemistry from Sun Yat-Sen (Zhongshan) University, and a PhD degree in chemistry from the University of Nebraska–Lincoln. After postdoctoral work on protein structure and engineering at the Brigham & Women's Hospital of Harvard Medical School and the Institute for Medicine and Engineering at the University of Pennsylvania, Dr. Xian became a lecturer at the University of Illinois at Urbana-Champaign, where she created a new biomaterials laboratory course for the Department of Materials Science and Engineering. Her research interests include protein engineering, tissue engineering, and antimicrobials.

1

Basic Laboratory Skills I

Experiments in this course aim to provide lab skill training in many disciplines. Before we begin, however, it is helpful to familiarize ourselves with some of the basic lab practices so that we can conduct the experiments more efficiently and safely. To get started, we will survey some of the most commonly used equipment and supplies found in our lab. We will also learn about proper waste disposal, an often ignored but extremely important practice that ensures protection of both the environment and ourselves. Next we will learn basic practices in liquid transfer and weighing; then, to integrate these practices, we will make two solutions that will be later used in our experiments. Last, we will review some of the basic practices in data processing.

I Commonly Used Lab Equipment and Supplies

- **Balances.** Top-loading electronic balances are commonly used nowadays. An electronic balance can be categorized as either *regular* or *analytical* depending on its weighing resolution, or "readability." A regular balance typically has a range of hundreds to thousands of grams and readability from milligrams to grams, while an analytical balance's range is typically of tens to hundreds of grams and its readability is ≤0.1 mg. When weighing samples, choose the right balance based on the weight of the sample and the accuracy requirement. A special type of mechanical balance, the trip balance, is also used in lab. (See Section IV for more details.)
- **Centrifuges.** There are a variety of centrifuges for different centrifugation needs, and one of the determining factors for centrifuge selection is the centrifugal force that is required for sedimentation of components in the sample. Typically, mini-centrifuges and microcentrifuges are used for quick processing of small volumes of samples (typically <2 ml each), whereas larger centrifuges, with either bench-top or floor models, accommodate larger volumes and offer more centrifugal power.
- **Glassware.** Glass beakers, Erlenmeyer flasks, test tubes, and bottles of different volumes are examples of typical glassware used in lab. To clean glassware, first wash with detergent until clean—water will flow as a sheet rather than streaks on clean glass surface—then rinse with water, and finally rinse with de-ionized water.
- **Gloves.** Disposable latex and nitrile gloves are the most commonly used in labs. Latex gloves are relatively inexpensive but have poor chemical resistance. Nitrile gloves have much better chemical resistance and should be worn when handling organic solvents or caustic reagents.
- **Liquid transfer.** Liquid transfer is handled in many different ways depending on the liquid volume and the accuracy requirement. Micropipettes, electronic

pipette-aids, transfer pipettes, and graduated cylinders are some of the most commonly used liquid transfer tools. (See Section III for more details.)

- **pH meter.** This is used for measuring the pH of solutions. Operate your pH meter according to the user's manual. It should be calibrated using pH standard buffers from time to time. *Important:* Do not allow the electrode to dry.
- **Mixing.** Vortex mixers, magnetic stirring plates, and shakers are examples of commonly used mixing equipment in the lab. In our experiments, we will also use specialized mixing equipment such as the sonicator and the homogenizer.
    - **Vortex mixer.** The high-speed vibration of a vortex mixer generates vortex in liquid, which results in quick mixing of liquid-liquid or liquid-solid. It is generally used for quick mixing in microcentrifuge tubes, test tubes, etc., in which vortexes can be generated.
    - **Magnetic stirring plate with stirring bar.** This is generally used for mixing in beakers, flasks, bottles, or other similar containers. For efficient stirring, the length of a magnetic stirring bar should be about 2/3 the diameter of the container's bottom.
    - **Shaker.** This is suitable for consistent mixing for prolonged periods of time. Some shakers are equipped with environment control such as water bath that can maintain constant temperature for incubating and shaking.
- **Refrigeration.** Some reagents need to be stored at low temperatures. For a given reagent, refer to the on-bottle label, the user's manual, or the material safety data sheet (MSDS) for the appropriate storage temperature.
- **Water purification.** Requirement of water purity varies depending on the application. Water purity is categorized as type I, II, and III, which typically requires the resistivity to be >18.2 mΩ·cm, >1 mΩ·cm, and >50 kΩ·cm, respectively; other parameters are restricted as well. For reagent preparation, type I water is generally preferred. On the other hand, for less demanding usage of purified water such as glassware rinsing, type III water is usually sufficient. (Note that de-ionized water can be corrosive to metal, thus it may not be suitable for certain applications.) De-ionization can be achieved by filtering water through mixed-bed ion-exchange resins that remove both anions and cations.
- **Disposable items.** Commonly used disposable items include 0.5-ml and 1.5-ml microcentrifuge tubes, pipette tips, serological pipettes (polystyrene), weighing boats, flint glass test tubes, etc. Observe waste disposal guidelines when discarding these items.

Exercises

1. Survey the centrifuges in your lab using Table 1.1 as an example.
2. Locate a vortex mixer, a magnetic stirring plate, and a shaker in your lab.
3. What type(s) of water purification equipment is(are) available in your lab? In your lab notebook, record the make and model of the equipment, the type of water (I, II, or III), and the resistivity threshold of the water purifier.

TABLE 1.1

A Survey of the Centrifuges in My Lab

Make	Model no.	Type[a]	Rotor	Max. RPM[b]	Max. RCF[c]	Vol. Capacity
Fisher Sci.[d]	—	Mini	Fixed-angle	6600	2200 ×g	6 × 2.0 ml
			Strip	6600	2200 ×g	8 × 0.2 ml or 16 × 0.2 ml
…	…	…	…	…	…	…

[a] Centrifuge type can be minicentrifuge, microcentrifuge, general-purpose centrifuge (bench-top or floor model), clinical centrifuge, etc.
[b] RPM: rotation per minute.
[c] RCF: Relative centrifugal force, expressed as a number of times of gravity, or "×g."
[d] Example (Fisher cat. no. 05-090-100). This model includes two rotors.

II Waste Disposal

It is important to observe applicable institutional guidelines when disposing of used supplies and reagents. Always ask your instructor if you are not sure how to dispose of a certain item. The following are some of commonly available waste disposals:

- **Broken glass disposal.** Clean, nonhazardous broken glassware, Pasteur pipettes, disposable glass tubes, etc. should be collected in a designated plastic-lined box clearly marked as "broken glass disposal." Residual chemicals or reagents should be removed from the glassware before disposal. *Example:* A Pasteur pipette was used for transferring chloroform. Before discarding it into broken glass disposal, leave the pipette in a fume hood until the residual chloroform has completely evaporated.
- **Regular trash.** Some disposable items such as plastic serological pipettes, plastic transfer pipettes, paper tissues, etc. that have not been used with toxic, caustic, or carcinogenic reagents may be discarded into regular trash. *Example:* A plastic transfer pipette was used for adjusting pH with HCl solution. Before disposing it into regular trash, rinse the pipette with water to remove any residual HCl solution.
- **Sharps disposal.** Sharps such as scalpels, blades, syringes, and needles need to be disposed of in special heavy-duty plastic sharps containers, which should be clearly marked. (Syringes, though made of plastic, are often required to be disposed of in sharps containers because of the association between syringes and needles.) *Example:* A scalpel was used for cutting up calf skin. When finished, wipe it on a piece of tissue paper to remove any stuck skin tissues and discard it into a sharps container.
- **Sink disposal.** Institutional and local municipal guidelines must be strictly followed for sink disposal. In general, only small amounts of nonhazardous inorganic salts, acids, and bases can be flushed down the sink with a large amount of water. *Example:* In our experiments, phosphate-buffered saline, electrophoresis running buffer, acetic acid solution, dilute HCl solution, NaOH solution, and excess salts such as NaCl taken from the bottles when weighing samples, etc. can generally be flushed down a sink with a generous amount of water.

- **Disposal of solvents, reagents, and chemicals.** These should be collected in specially designated containers and disposed of following proper guidelines. *Example:* 70% ethanol is used for disinfecting calf skin before collagen extraction. When finished, the 70% ethanol should be collected in a container clearly labeled with "70% ethanol waste with trace calf skin" or similar wording.

Exercise

Locate the broken glass disposal and sharps disposal containers in your lab.

III Liquid Transfer

III.1 Micropipettes

Micropipettes, also called micropipetters or pipettes, are piston-driven air-displacement devices that are typically used to transfer liquid volumes in the range of 1 µl to 1000 µl with high precision and accuracy, which can be as low as <0.5% or as high as 5% depending on the micropipette and the pipetted volume. Specialized micropipettes are available for larger or smaller volumes.

III.1.1 Anatomy of a Micropipette

Different micropipettes may have different designs, but they all have certain common features and are operated in similar ways:

- The handle. Grab the handle firmly with your hand.
- A push button, or plunger, on the top of the micropipette. Use the thumb to push the plunger for delivery of the liquid. There should be two stops when pushing down on the plunger. Usually, for delivery of the bulk volume, press the plunger to the first stop and to deliver the residual liquid, press the plunger to the second stop.
- A volume adjustment dial and a numeric display. Turn the dial to set the volume of the liquid to be delivered. For accuracy, turn the dial past the desired setting, and then turn it back down gradually to the correct setting. The unit for the set volumes is µl; make sure that the decimal place is located correctly on the dial.
- A shaft that leads to the tip cone. To attach a tip to the tip cone, firmly push down the pipette (but avoid using excessive force).
- A tip ejector. To eject a tip, point the tip to the waste receptacle and press the ejector with your thumb.

III.1.2 Pipetting Techniques

Observe the following guidelines when using a micropipette:

- Pick the right micropipette. Do not use a micropipette outside its designated range; otherwise it could be damaged mechanically. When a volume to be pipetted is within the ranges of different micropipettes, use the one with the smallest range.

For example, to pipette a volume of 18.2 µl, use a 20-µl micropipette instead of a 100-µl micropipette.

- *Never* allow liquid to get inside the micropipette barrel. This will lead to cross-contamination between samples, and potential damage to the micropipette as well. To avoid this situation, pipette liquid in smoothly, and do not tilt the micropipette too far from vertical when it is holding liquid.
- Press and release the plunger smoothly. Do not allow it to snap back.
- When loading the pipette tip onto the tip cone, make sure that the tip is firmly attached but avoid using excessive force—a light tap or two on the tip rack will usually do.
- Pipetting in: Submerge the pipette tip slightly (several mm) beneath the surface of the liquid. Do not plunge the tip into the liquid.
- Pipetting out: Hold the tip against the inner wall of the receptacle for a steady and smooth delivery of the liquid.

Select the following pipetting techniques based on the volume of the liquid to be transferred, and whether it is nonviscous, viscous, or foamy:

- **The forward technique.** The forward technique is typically used for nonviscous liquid. To pipette using the forward technique, press the plunger to the first stop and release to draw liquid in through the tip. Wipe the tip against the liquid container to remove excess liquid on the outside of the tip. In the new receptacle, press the plunger to the first stop to deliver the liquid. Wait for a second or two, and then press the plunger to the second stop to empty any residual liquid. Release the plunger and eject the tip.
- **The reverse technique.** The reverse technique is suitable for viscous or foamy liquid, or very small volume. Examples of viscous or foamy liquids include glycerol solutions, protein solutions, and detergent solutions. To pipette using the reverse technique, first press the plunger all the way to the second stop, then release slowly to draw liquid in. Again, wipe the tip against the liquid container to remove excess liquid on the outside of the tip. To deliver, press the plunger to the first stop and hold for one or two seconds. The remaining liquid should be released back into the original liquid container or discarded.
- **The pumping technique.** Some liquids, such as whole blood, are viscous and tend to cling to pipette tips. Repeated pumping is necessary to deliver the full volume. To pipette using the pumping technique, first press the plunger to the first stop, then release to draw up liquid. Press the plunger to the first stop to deliver the liquid, and then release it smoothly. Repeat the pumping motion until all the liquid inside the tip is delivered. To finish, press the plunger to the second stop to deliver any residual liquid.

III.1.3 Before You Put Away the Micropipette ...

After work is finished, do the following before you put the micropipette away for the day:

- Clean the micropipette. Wipe away any moisture or soiled spots. If the micropipette has been used to handle hazardous materials, make sure that no residue is left behind.

- Turn the volume adjustment to the highest setting. This is to prevent the spring inside the micropipette from being compressed for a long period of time, which helps to maintain the accuracy of micropipette.
- For small-volume micropipettes (<10 μl), it is recommended that they be stored with pipette tips attached, so that in case the micropipettes are dropped, their small tip cones are protected.

III.2 Pipet-Aids

A pipet-aid is a device that is used with a serological pipette to transfer liquid volumes from <1 ml to ~100 ml. Commonly used pipet-aids are electrical and are controlled by two buttons: Press the top button to draw up liquid and the down button to deliver the liquid. A serological pipette is attached to the mouthpiece of a pipet-aid.

Pay attention to the following when using a pipet-aid for our experiments:

- ***Caution:*** Do *not* draw liquid into the mouthpiece of the pipet-aid.
- Pipetting organic solvent: To pipette organic solvents, glass serological pipettes *must* be used. Polystyrene pipettes will dissolve in many organic solvents. To tell a glass pipette from a polystyrene pipette: 1) read the label on the pipette; 2) feel the pipette: glass pipettes are heavier and clink when bounced on a hard surface. Pay attention to whether the glass pipette is reusable or disposable.
- Pipetting in: Liquid should be drawn up smoothly; do not allow it to "spring" (like spring water gushing out of the ground) through the opening.
- Pipetting out: Rest the tip of the serological pipette against the inner wall of the receptacle if possible, and deliver the liquid smoothly.

III.3 Pasteur Pipettes

Pasteur pipettes are glass pipettes with long, fine tips that need to be attached to rubber bulbs for pipetting. These pipettes are inexpensive and have excellent chemical resistance; they are especially suitable for transferring organic solvents and reagents when volume accuracy is not required. ***Caution:*** Pasteur pipettes should be disposed of in broken glass disposal.

III.4 Plastic Transfer Pipettes

These are disposable plastic pipettes with "built-in" suction bulbs. They can be used to transfer small volumes of liquid (<10 ml each transfer) when volume accuracy is not required. (They are also very useful for other tasks; for example, a transfer pipette can be used for resuspending a pellet after centrifugation since it can be used as a stirring rod as well as a pipette.)

III.5 Graduated Cylinders and Volumetric Flasks

Graduated cylinders and volumetric flasks are liquid-measuring devices that are mainly used for making solutions. Volumetric flasks are used when the accuracy requirement for concentration is very stringent. (For the experiments in this course, graduated cylinders are sufficient for making solutions. See Section V.)

IV Weighing

IV.1 Electronic Balance

Weighing with top-loading electronic balance is straightforward: Place a weighing boat or weighing paper on the platen, tare (zero) the balance, add the sample to the weighing boat or paper, and read the displayed weight. Observe the following guidelines when weighing solid samples:

- **Range:** Each balance has its range and accuracy. Do not use a balance outside its range, and pick a balance with the right accuracy.
- **For analytical balance:** It is important for an analytical balance to be leveled. Check the window on the balance in which an air bubble is sealed in water: if the bubble is not centered, adjust the screws on the feet of the balance until it is centered.
- **Weighing powder sample:** When weighing powder, use spatula to transfer the powder to the weighing vessel; when it is getting close to the target weight, lightly tap on the wrist of the hand that is holding the spatula and allow a small amount of powder to fall into the weighing vessel, and repeat until the target weight is reached.
- **Cleaning up:** Use a tissue or brush to clean up any spilled sample. Tare the balance.

IV.2 Trip Balance

Trip balance is a type of mechanical balance that is often used for measuring the *relative* weight of one sample against another. It is particularly useful when weighing samples for centrifugation, where two centrifuge tubes across from each other in the rotor must be equal in weight. To balance two centrifuge tubes, place two tube holders (such as 100-ml beakers) on the two platens, and then zero the needle in the middle by sliding the weight on the balance beam. Next, place one centrifuge tube in each holder; add or subtract samples until the needle is zeroed again. If the two tubes are holding the same sample, then use a transfer pipette to transfer the sample back and forth until the two tubes are balanced. Remember to include the caps of the centrifuge tubes for the balancing. *Note:* Balancing with a trip balance is a must for high-speed centrifuges, where a small difference in the weight is augmented by the high centrifugal force. For minicentrifuge or microcentrifuge that uses microcentrifuge tubes, balancing is achieved by weighing the microcentrifuge tubes with electronic balance and adjusting the liquid volumes accordingly. For short centrifugation time at low speed (<1000 xg), sometimes "eyeball balancing" is good enough.

Exercises

1. Survey the balances in your lab using Table 1.2 as an example.
2. Add water into two 50-ml centrifuge tubes to roughly 2/3 full. Balance the two tubes on a trip balance.

TABLE 1.2

A Survey of the Balances in My Lab

Make	Model No.	Type[a]	Range (g)	Readability (g/mg)
Ohaus Scout Pro[b]	SP601	Top loading, general	0–600	0.1 g
…	…	…	…	…

[a] Types of balances include general, analytical, top loading, trip balance, etc.
[b] Fisher cat. no. 01-921-13.

V Making Solutions

V.1 Phosphate Buffered Saline

In the following exercise, you will make 500 ml of phosphate-buffered saline (PBS) with the following composition. This solution will be used in Module I (see Chapter 2).

- 155 *mM* NaCl
- 1.0 *mM* KH_2PO_4
- 3.0 *mM* Na_2HPO_4

Procedures

1. *Check-in*
   a. Reagents
      - NaCl, solid
      - KH_2PO_4, solid
      - Na_2HPO_4, solid
      - De-ionized water
      - 1.0 *M* HCl solution
      - 1.0 *M* NaOH solution
   b. Equipment and supplies
      - Balance, with weighing boats and spatula
      - Kimwipe tissue paper
      - Graduated cylinders, 500 ml and 100 ml
      - Beaker, 500 ml or 600 ml
      - 500-ml bottle, glass or plastic
      - Magnetic stirring plate and stir bar
      - pH meter
      - Wash bottle with de-ionized water
      - Labeling tapes

TABLE 1.3

Components of PBS

Component	Concentration (*mM*)	Mol. Formula	Formula Weight	Weight Needed for 500 ml PBS
NaCl	155			
KH_2PO_4	1.0			
Na_2HPO_4	3.0			

2. *Calculations:* Note that some salts are available as either anhydrites or hydrates. Check the labels on the bottles for NaCl, KH_2PO_4, and Na_2HPO_4 for the molecular formula of the salts and their formula weights. Calculate the weight of each salt needed for the PBS. Organize your data in Table 1.3.
3. *Weighing samples:* Use an appropriate balance and proper techniques to weigh NaCl, KH_2PO_4, and Na_2HPO_4 according to Table 1.3. Wipe the spatula clean with a Kimwipe tissue before using it for the next chemical, but use a different weighing boat for each chemical. Add the samples to a 500-ml or 600-ml beaker.
4. *Dissolving the salts:* Add a stir bar of an appropriate size and ~400 ml of de-ionized water to the beaker. When adding water, position the spout of the graduated cylinder so that the water flows down along the inner wall of the beaker. Do not let the water splash. Place the beaker on a magnetic stirring plate and stir the solution with moderate speed until the salts are completely dissolved.
5. *pH adjusting:* Turn on the pH meter, which has already been calibrated by your instructor. Take the electrode out of the storage buffer, open the air inlet (usually located at the upper end of the electrode), and rinse the lower end of the electrode with de-ionized water from a wash bottle. Position the tip of the electrode about 2–3 cm below the surface of the solution and read the pH meter. If it is >7.4, use a plastic transfer pipette to add 1.0 *M* HCl solution drop by drop to adjust it to 7.4; if the pH is <7.4, use 1.0 *M* NaOH solution in the same manner. When finished, rinse the electrode with de-ionized water again, blot the water off with a Kimwipe tissue, and place it back into the storage buffer.
6. *Adjusting the volume:* Pour the solution from the beaker to a 500-ml graduated cylinder. Rinse the beaker with de-ionized water several times and add the rinse to the graduated cylinder as well. Finally, add de-ionized water to the graduated cylinder so that the volume is exactly 500 ml. (Change in pH due to change of the solution volume is assumed to be small. Measure and adjust the pH again if necessary.) Pour the PBS into a 500-ml bottle.
7. *Labeling:* Label the bottle with the name of the solution (PBS), the pH, your group's name, and the date. You may also write the composition of the solution on the label for your own convenience. *Important:* Always label your samples and reagents with at least the names of the chemicals or materials, your (group's) name, and the date; other pertinent information such as the pH of a buffer, the toxicity if any, etc. should be added to the label as well.
8. *Finishing up:* Store the PBS. Wash and rinse the beaker, the graduated cylinders, the stir bar, and the spatula. The used weighing boats and transfer pipettes can be discarded into regular trash after rinsing with tap water.

V.2 Acetic Acid Solution (0.50 *M*)

In the following exercise, you will make 200 ml of 0.50 *M* acetic acid solution, which you will use later in Module II (see Chapter 3).

Procedures

1. *Check-in*
   a. Reagents
      - Glacial acetic acid (pure acetic acid)
      - De-ionized water
   b. Equipment and supplies
      - 200-ml graduated cylinder
      - Glass bottle, ≥200 ml
      - Pipet-aid
      - Serological pipette, polystyrene and glass
      - Fume hood
2. *Calculations:* Glacial acetic acid is in liquid form at room temperature, and its density is 1.049 g/ml. The molecular weight of acetic acid is 60.05 g/mol. Calculate the molarity of glacial acetic acid at room temperature, and then calculate the volume of glacial acetic acid, V_{gaa} in ml, that you need to make 200 ml of 0.50 *M* acetic acid solution.
3. *Making the solution:*
   a. Measure 200 ml de-ionized water with a graduated cylinder and add it to a glass bottle.
   b. With a pipet-aid and a plastic serological pipette, remove V_{gaa} ml of water from the bottle.
   c. In a fume hood, pipette V_{gaa} ml of glacial acetic acid with a dry glass pipette and add it to the de-ionized water. (Remember: When diluting acid, always add acid to water instead of adding water to acid.) Pipette up and down several times for a complete delivery of the acetic acid into the water. Cap the bottles. Swirl the acetic acid solution to mix evenly.
   d. Label the acetic acid solution properly. (See V.1, Step 7.)
4. *Finishing up:* Store the 0.50 *M* acetic acid solution. Rinse the glass pipette with water and de-ionized water. Let the pipettes and the graduated cylinder dry. (The polystyrene pipette may be reused since it was only used for water.)

VI Error Analysis

It is assumed that you have learned error analysis in your previous studies. A paper by Cumming et al. (2007) offers an excellent review with examples. The following exercises are designed as a review and refreshment for the error analysis methods that will be used in this course.

TABLE 1.4

What is the Number of Significant Figures for the Following Measurements?

Parameter	Value (unit)	No. of Sig. Figs.
Density	1.049 g/ml	
Absorbance	0.330	
Readability	0.0001 g	
Volume	1000.0 µl	

VI.1 Significant Figures

Significant figures (or significant digits) indicate the degree of error for a given measurement. For example, a concentration given as 0.50 *M* indicates possible error of 0.01 *M*, whereas a concentration given as 0.5 *M* indicates possible error of 0.1 *M*. In multiplication and division, the number of significant figures in the result should be the same as the smallest number of significant figures among the multiplied or divided quantities; and in addition and subtraction, the result should have the same number of decimal places as the added or subtracted quantity with the fewest number of decimal places.

Exercises

1. Write down the number of significant figures for the measurements in Table 1.4.
2. There is a bottle of commercial NaOH solution with a concentration of 1.000 *N* in your lab. If you accidentally mix it with a bottle of homemade 1.0 *N* NaOH solution, how should you label the concentration of the mixed solution?
3. Check your calculation in Section V.2, Step 2, for the volume of glacial acetic acid needed to make 200 ml of 0.50 *M* acetic acid solution. What should the number of significant figures be in your result?
4. Check your calculation results in Table 1.3 and make sure that the correct numbers of significant figures are used for each results.

VI.2 Accuracy and Precision

Accuracy is defined by how closely a measurement agrees with the true or accepted value, and it reflects the system errors in the measurement. Precision is defined by the reproducibility of the measurement, and it reflects the random errors in the measurement. To evaluate the accuracy and precision of a set of independent measurements, two parameters are frequently used: the mean (M) and the standard deviation (SD). The mean is defined as

$$\mathrm{M} = \frac{1}{N}\sum_{i=1}^{N} X_i \qquad \text{(BLS I.1)}$$

where N is the number of measurements, and X_i is the result of measurement number i.

The standard deviation is defined as

$$SD = \sqrt{\frac{\sum_{i=1}^{N}(X_i - M)^2}{N-1}} \qquad \text{(BLS I.2)}$$

Note that when reporting the measurement result, the significant figures of the mean should reflect the standard deviation; a measurement is no more accurate than the errors in the measurement would allow. For example, if the mean of two concentration measurements for a protein solution is 0.4571 *mM*, but the standard deviation is 0.054, then the concentration of the protein should be reported as 0.46 ± 0.05 *mM* (or 0.457 ± 0.054 *mM*).

In the following exercise, you will perform a calibration experiment using your pipetting and weighing skills, and then you will perform error analyses on the data to determine the accuracy and precision of a micropipette.

Exercises

1. *Check-in*
   a. Reagents
      - De-ionized water
   b. Equipment and supplies
      - 100-µl or 200-µl micropipette
      - Analytical balance
      - 0.5-ml or 1.5-ml microcentrifuge tubes
2. *Preparation:* Press a 0.5-ml or 1.5-ml microcentrifuge tube into a block of Styrofoam so that the tube is standing, and place the block on the platen of the analytical balance. Tare the balance.
3. *Measurement*
   a. The outline: With an analytical balance, measure the weight of a given volume of water delivered by the micropipette to determine the accuracy of the micropipette; for each volume, multiple measurements are made to determine the precision of the micropipettes. (Here we assume that the analytical balance is accurate.) Volumes are to be measured according to Table 1.5.
   b. Add the first volume of de-ionized water into the microcentrifuge tube using the 100-µl or 200-µl micropipette according to Table 1.5, and take a reading of the analytical balance once it has stabilized. Tare the balance, and add the next volume, take a reading, and so on. Replace the 0.5-ml tube if it becomes full. Collect your data in Table 1.5.
4. *Data processing*
   a. Calculate the mean and standard deviation for each volume in Table 1.5. Take care to use the correct number of significant figures for the mean so that it is in agreement with the standard deviation.
   b. Using the data in Table 1.5, plot the weights vs. the volumes. Add error bars to the plot according to the standard deviations.
   c. Fit the plot with a linear function.

TABLE 1.5

Measuring the Accuracy and Precision of a 100-μl Micropipette

Volume	Weight (mg)	Mean (mg)	SD
10.0 μl	Measurement #1:		
	Measurement #2:		
	Measurement #3:		
25.0 μl	Measurement #1:		
	Measurement #2:		
	Measurement #3:		
50.0 μl	Measurement #1:		
	Measurement #2:		
	Measurement #3:		
75.0 μl	Measurement #1:		
	Measurement #2:		
	Measurement #3:		
100.0 μl	Measurement #1:		
	Measurement #2:		
	Measurement #3:		

Discussion

1. According to the data in Table 1.5, do the accuracy and the precision remain constant for different volumes?
2. Why is it that if a volume to be pipetted fits the ranges of different micropipettes, then we should pick the micropipette with the smallest range (Section III.1.)?
3. What is the slope of the linear fit for the weight vs. volume plot from Table 1.5? Is it what you expected?
4. For the plot in Question 3, does the slope of the linear fit reflect the accuracy or the precision of the measurements? Does the correlation coefficient of the linear fit (R) reflect the accuracy or the precision of the measurements?
5. If you are trying to establish a linear calibration curve, and your data look like the ones in Figure 1.1, should you include data point #5 in the linear fitting?

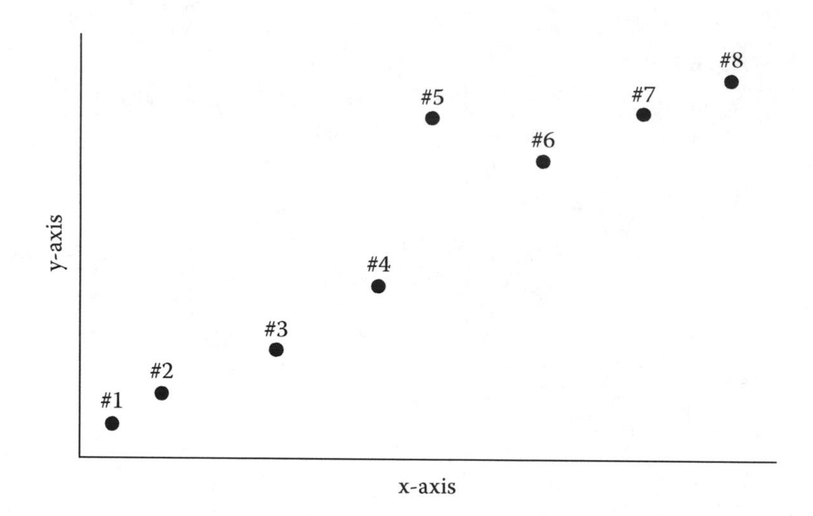

FIGURE 1.1
Hypothetical data for a linear calibration curve.

Reference

Cumming, G., F. Fidler, and D.L. Vaux, Error bars in experimental biology. *J. Cell. Biol.*, 177, 7–11, 2007.

2

Module I. Drug Delivery: Controlled Release of Encapsulated Protein from PLGA Microspheres

The need for controlled release of drugs is obvious. Conventional drug delivery methods such as subcutaneous injection and oral ingestion have many problems, including side effects due to non-site-specific delivery of the drug, drug concentrations being too high or too low, patient compliance due to strict drug intake schedule, and so forth. An ideal drug delivery system is one that releases the drug to a specific location at a constant rate and at a desirable concentration of the drug (Figure 2.1).

Many drug delivery systems have been designed for this purpose (Ratner et al. 2004), and among them are the biodegradable monolithic systems. In these systems, the drug is dispersed, or encapsulated, in a matrix that is usually composed of biodegradable polymer, and drug release is dependent on the biodegradation of the matrix. Two different degradation mechanisms have been defined (Figure 2.2): one is bulk erosion, which is akin to a gradual shattering of the matrix, and the other is surface erosion, which is like gradually shaving off the surface layer by layer. A variety of molecules, from small-molecule

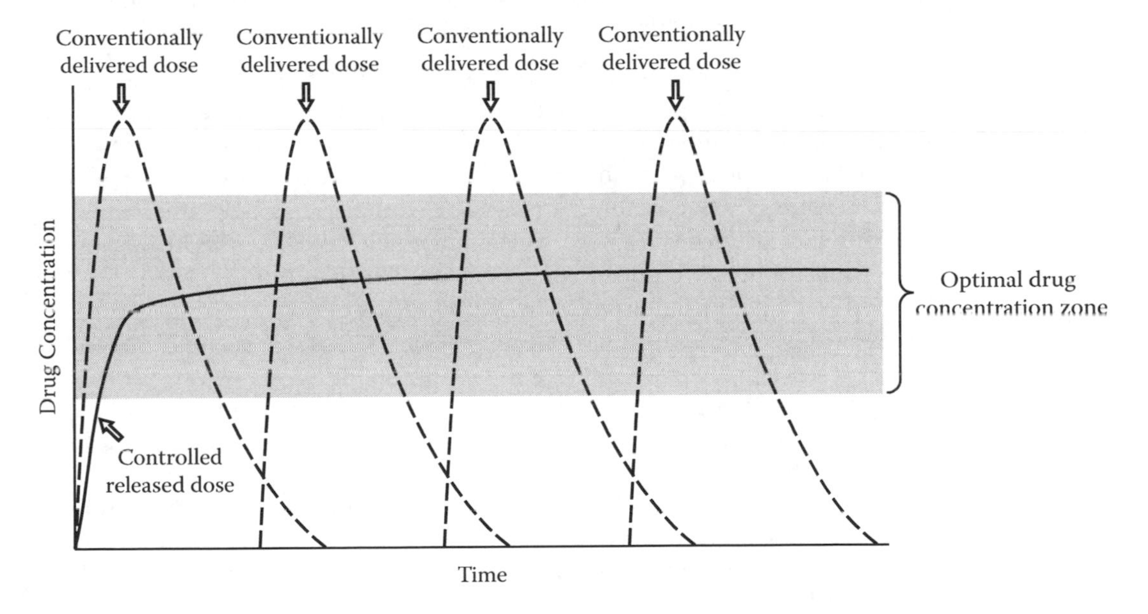

FIGURE 2.1
Controlled release drug delivery compared to conventional drug delivery. For conventional drug delivery systems such as subcutaneous injection or oral ingestion (dotted line), drug doses are taken at regular intervals to maintain the presence of the drug, but the drug concentration is not always optimal. In ideal controlled release systems, the presence and concentration of the drugs can be maintained for long periods of time, which may be hours, days, months, or even years.

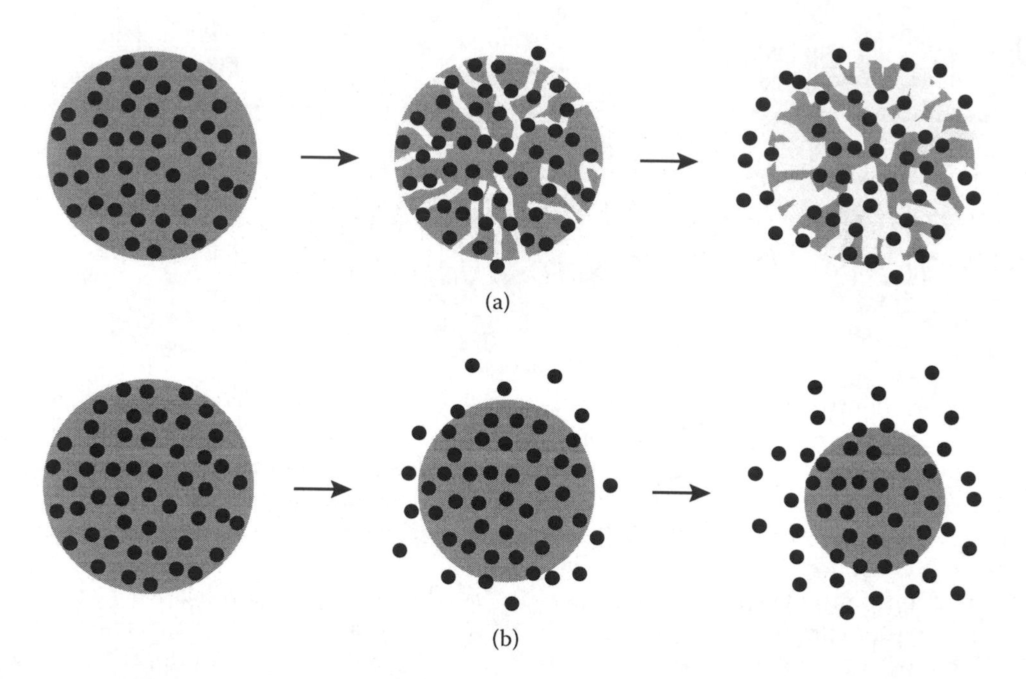

FIGURE 2.2
Controlled release of encapsulated drug from monolithic matrix through two different degradation mechanisms. (a) Bulk erosion, in which cracks and channels form over time that facilitate the release of the drug. A characteristic of this release mechanism is the burst release of the drug at the final stage when the matrix quickly and completely disintegrates. (b) Surface erosion, in which degradation is restricted to the surface. This mode is more desirable since the drug release rate is close to being constant.

drugs to large proteins such as antibodies, have been encapsulated in monolithic drug delivery devices for constant-dosage delivery over various periods of time. An example of such device is Arestin® (OraPharma, Inc., Warminster, Pennsylvania), a commercial name for minocycline microspheres in 1 mg dosages. These microspheres can be injected into the periodontal pockets after scaling and root planing (SRP). The microspheres adhere to the surrounding gum tissues, and as the microspheres degrade, minocycline, a broad-spectrum antibiotic that kills infectious bacteria, is constantly released into the periodontal fluid and helps the gum tissues to heal around the teeth.

In this module, we will focus our studies on how to use biodegradable monolithic drug delivery systems to release proteins as therapeutic reagents. Why are we interested in proteins? On one hand, proteins participate directly or indirectly in almost all molecular activities in our bodies, thus the potential of using proteins to remedy, enhance, or inhibit these activities is enormous. On the other hand, the labile nature of protein molecules poses unique challenges and requires special strategies for the encapsulation process (Putney 1998). We will learn about the rationales behind the protein encapsulation strategies, and understand how these strategies can be improved to maximize the potential of proteins as therapeutic drugs.

In our experiments, we will fabricate microspheres of poly(lactic-*co*-glycolic acid) (PLGA) with encapsulated bovine serum albumin (BSA), and then study the release of BSA through *in vitro* biodegradation of the microspheres. PLGA is a biodegradable copolymer of lactide and glycolide that has been approved for drug delivery by the U.S. Food and Drug Administration (FDA). At the molecular level, the final products of PLGA degradation are

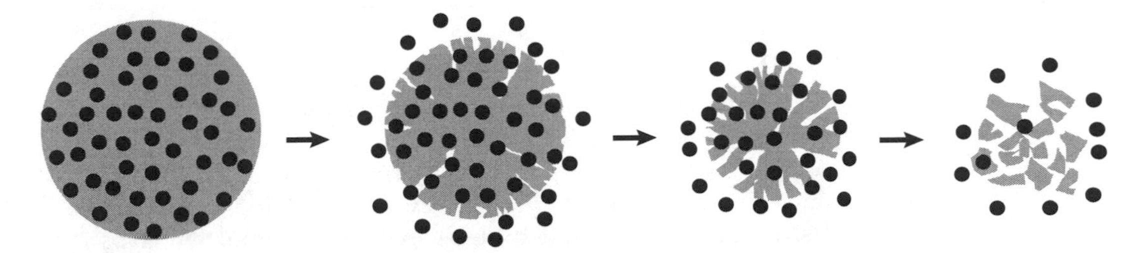

FIGURE 2.3
The drug release mechanism for PLGA microspheres is a combination of surface erosion and bulk erosion. Surface erosion occurs as the microspheres come in contact with the release medium; as water-filled channels start to form and subsequently enlarge, drug release is enhanced. Eventually, the microspheres undergo catastrophic collapse and completely disintegrate.

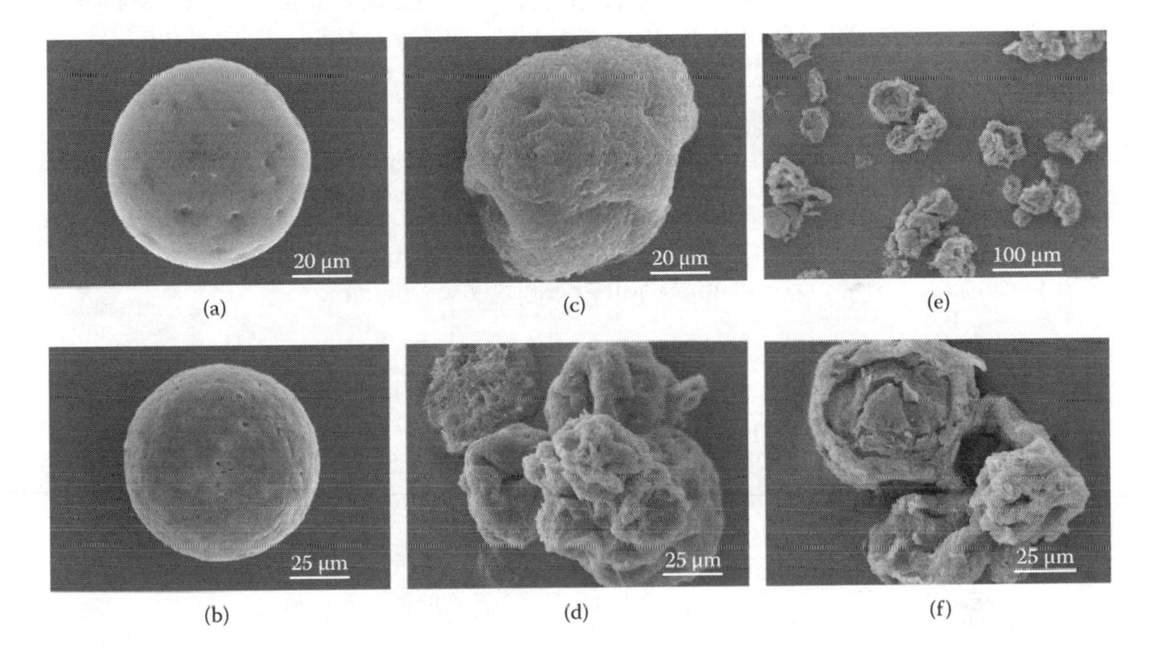

FIGURE 2.4
Scanning electron microscopy (SEM) Images of PLGA 50/50 microspheres at different degradation stages: (a) 24 hours, (b) 3 weeks, (c) 5 weeks, (d) 7 weeks, and (e, f) 11 weeks. The microspheres are prepared with water-in-oil-in-water double emulsion with encapsulated BSA or human parathyroid hormone (PTH). (Reprinted from Wei Guobao, Glenda J. Pettway, Laurie K. McCauley, and Peter X. Ma, "The release profiles and bioactivity of parathyroid hormone from poly(lactic-*co*-glycolic acid) microspheres," *Biomaterials*, 25, 345–352, 2004, with permission from Elsevier.)

lactic acid and glycolic acid, both natural metabolic products in the human body, which is a main factor that contributes to the excellent biocompatibility of PLGA. At the macroscopic level, the degradation of PLGA microspheres is a combination of bulk erosion and surface erosion (Figure 2.3 and Figure 2.4), and many factors, like the molecular weight of the PLGA, the size of the microspheres, and the porosity of the matrix, can affect the protein release process. The protein to be encapsulated, BSA, is a protein found in bovine blood plasma. Because of its relative stability and abundant availability, BSA is often used as a model protein for studies on protein encapsulation.

Session 1. Encapsulation of BSA in PLGA Microspheres

The essence of protein encapsulation is the even dispersion of the protein into the polymer matrix, and there are several methods that serve this purpose. We will use a *water-in-oil-in-water* double emulsions procedure. BSA, like most proteins, is soluble in aqueous solution but not in organic solvent; on the other hand, PLGA is soluble only in relatively nonpolar organic solvents. To disperse BSA into a PLGA matrix, two emulsion steps will be used: first, aqueous solution of BSA is dispersed in PLGA-dichloromethane solution by sonication, which creates a *water-in-oil* emulsion—we also call it the first-degree emulsion (Figure 2.5a); next, the first-degree emulsion is mixed with polyvinyl alcohol (PVA) solution and emulsified by homogenization, which creates the second-degree emulsion, essentially an *oil-in-water* emulsion (Figure 2.5b). Here, PVA acts as a surfactant at the oil-water interface and helps to stabilize the emulsion. The result of these two emulsification steps is the encapsulation of BSA solution droplets in microspheres of PLGA solution. The liquid PLGA microspheres become hardened when the dichloromethane evaporates, and the encapsulation is complete when water is removed from the microspheres through freeze-drying.

PLGA comes in many varieties, and they are different in the lactide to glycolide ratio, molecular weight, resorption time, and so forth, so we need to pay attention to the chemical and physical characteristics of a given batch of PLGA provided by the manufacturer. (Considering the duration of our course, a resorption time of 1–2 months would be ideal. A 50/50 copolymer of *d-/1*-lactide and glycolide with a molecular weight of ~70,000 Daltons usually satisfies this resorption time requirement.)

Safety Note

An organic solvent, dichloromethane, is used in the following experiment. Wear nitrile gloves that are resistant to dichloromethane. Do ***not*** use latex gloves. All procedures involving dichloromethane should be carried out in a fume hood.

Procedures

Part 1 (Day 1). The Double Emulsions

1. *Check-in*
   a. Samples and materials
      - PLGA pellets
   b. Reagents
      - Dichloromethane
      - BSA, solid
      - 9% PVA solution
      - 0.01x phosphate-buffered saline (PBS) (composition: 1.55 *mM* NaCl, 0.01 *mM* KH_2PO_4, and 0.03 *mM* Na_2HPO_4)
   c. Special equipment and supplies*
      - 20-ml glass vial
      - Glass serological pipettes

* This refers to equipment and supplies that are used for specific procedures besides the general lab equipment and supplies described in Chapter 1: Basic Lab Practices I.

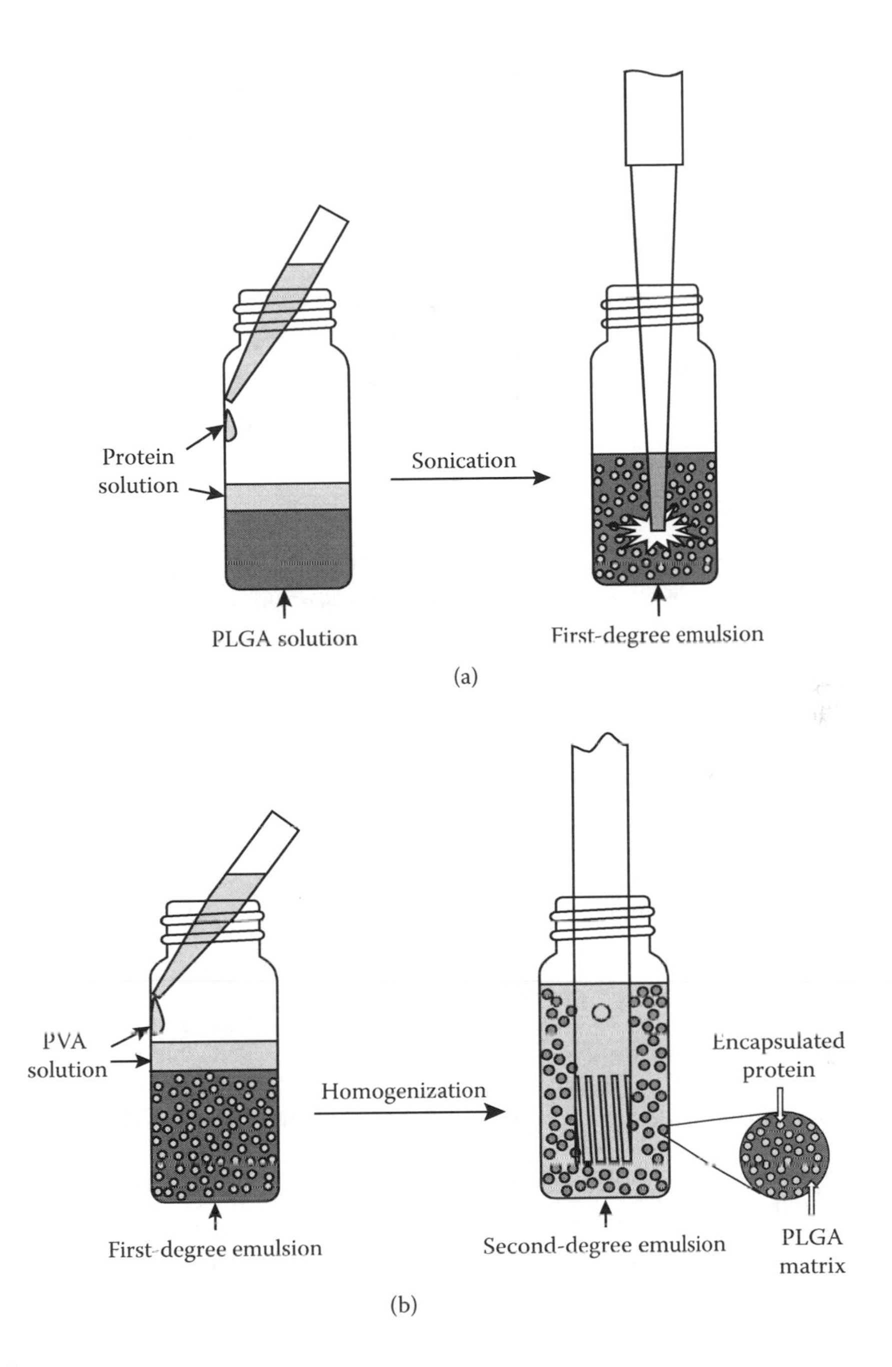

FIGURE 2.5
Fabrication of PLGA microspheres with encapsulated BSA using a double-emulsion process. (a) The first-degree emulsion: BSA solution is added to the PLGA solution, and a sonication probe is used to break up the BSA solution into tiny droplets that are evenly distributed in the organic solvent, creating a water-in-oil emulsion. (b) The second-degree emulsion: PVA solution is added to the first-degree emulsion, and a homogenizer is used to disperse the organic phase into small droplets, creating an oil-in-water emulsion. Thus, water-in-oil-in-water double emulsions is generated.

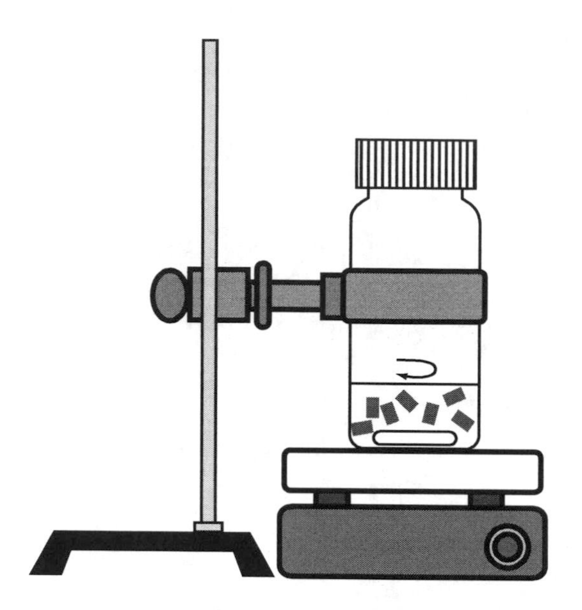

FIGURE 2.6
A setup to dissolve PLGA pellets in dichloromethane. A glass vial is held in place with a clamp and placed on top of a magnetic stirring plate. PLGA pellets are dissolved in organic solvent with stirring.

- Stand and clamps
- Sonicator with microtip
- Homogenizer

2. *Setup:* Place a magnetic stirring plate in a fume hood. Center and secure the 20-ml glass vial on the stirring plate with a small stirring bar inside (Figure 2.6).
3. *Dissolving PLGA:* Take note of the size, color, and opacity of the PLGA pellets. Weigh ~200 mg of PLGA pellets on an *analytical* balance. Record the weight to 0.1 mg. Using a *glass* serological pipette, add 5 ml of dichloromethane to the 20-ml glass vial. Start stirring at a medium speed. Add the PLGA pellets one by one into the glass vial till all is added. Cap the vial and continue to stir the pellets. It will take 15–30 minutes for the pellets to completely dissolve. Meanwhile, proceed to the next step.
4. *Preparing BSA Solution:* Weigh ~50 mg BSA solid in a 1.5-ml microcentrifuge tube. Add 1000 µl of 0.01× PBS to the microcentrifuge tube, and then pipette up and down to completely dissolve the protein. ***Caution:*** Pipette slowly to minimize foaming.
5. *First-degree emulsion*
   a. Make sure that both the PLGA solution and the BSA solution are ready. Insert the microtip of the sonicator inside the 20-ml glass vial. *Important:* The microtip should be close to the bottom of the vial for maximum efficiency, but it should not touch the vial (Figure 2.5a). Set the sonicator to a power output of ~40 watts.
   b. Add 500 µl of BSA solution to the glass vial. Sonicate according to the following sequence: **on** for 5 seconds—**off** for 10 seconds—**on** for 5 seconds. Observe the mixture before and after sonication.

   c. Remove the microtip and immediately proceed to the next step.

6. *Second-degree emulsion*
   a. Have the 9% PVA solution ready. Insert the generator probe of the homogenizer to the middle of the 20-ml glass vial.
   b. Pipette 10 ml of the 9% PVA solution into the glass vial and make sure that the liquid level is above the hole in the generator probe (Figure 2.5b).
   c. Turn on the homogenizer, gradually turn up the power until mixing takes place, and let it run for 15 minutes. Adjust the power and the probe position if necessary, but make sure that the hole is always submerged.
7. *Solvent evaporation*
   a. Add 20 ml of the 9% PVA solution to a 100-ml Erlenmeyer flask and a stirring bar. Turn on the stirring plate so that the PVA solution is stirred steadily, then add the second-degree emulsion from the glass vial to the PVA solution using a *glass* transfer (Pasteur) pipette.
   b. To completely transfer the second-degree emulsion, rinse the glass vial by adding 15 ml of the 9% PVA solution to submerge the hole of the generator probe and running the homogenizer for a few seconds. Transfer the rinse to the 100-ml Erlenmeyer flask. Repeat the rinse once more.
   c. Leave the mixture stirring overnight; make sure that the stirring is stable. Do not seal or cover the flask.
   d. After solvent evaporation is complete, your instructor will seal the flask and store it at 4°C.

Part 2 (Day 2). Harvesting the Microspheres

1. *Check-in*
   a. Samples and materials
      - Your group's PLGA microsphere suspension from Part 1.
   b. Reagents
      - De-ionized water
   c. Special equipment and supplies
      - Round-bottom centrifuge tubes and necessary adaptors, 80 ml
      - Disposable conical centrifuge tube, 50 ml
      - Liquid nitrogen
      - Centrifuge with fixed-angle rotor
      - Freeze-dryer
      - Optical microscope with image capture
      - Microscope slides and coverslips
      - Microscope stage micrometer
2. *Harvesting the microspheres*
   a. Swirl the BSA-PLGA microsphere suspension to resuspend the sedimented microspheres. Distribute the suspension equally to two 80-ml round-bottom centrifuge tubes. Rinse the flask with ~10 ml de-ionized water and distribute

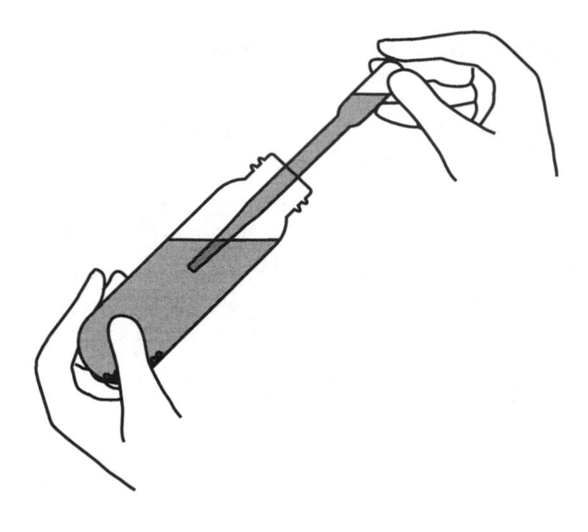

FIGURE 2.7
Removing the supernatant after centrifugation. PLGA microspheres form a potentially loose pellet after centrifugation; therefore, the pellet should be positioned to the bottom of the centrifuge tube, and a transfer pipette or a serological pipette is used to remove the supernatant. (*Note:* If the pellet is tightly packed and firmly attached to the tube wall, then it is possible to position the pellet on top of the supernatant; the supernatant can then be pipetted out or simply poured out.)

the rinse to the centrifuge tubes as well. Afterwards, add de-ionized water to the fill line of the centrifuge tubes.

b. Balance the centrifuge tubes. Cap the tubes and mix the content well by flipping them several times. Centrifuge the suspension for 10 minutes at 10,000 rpm (or >12,000 ×g).

c. Hold the centrifuge tube so that the pellet is at the bottom of the tube (Figure 2.7). Decant or pipette off the supernatant. Apply caution when pipetting since the pellet may be loose.

3. *Washing the PLGA microspheres*

   a. Pour de-ionized water to each centrifugation tube to >60% of the tube capacity. Pipette up and down to completely resuspend the pellets.

   b. Balance the tubes and centrifuge for 10 minutes at 10,000 rpm (or >12,000 ×g). Discard the supernatant.

   c. Repeat the last two steps once more.

4. *Freeze-drying the PLGA microspheres*

   a. Resuspend each pellet with 5 ml of de-ionized water. To save a sample of the microspheres, pipette 50 μl of the microsphere suspension to a 1.5-ml or 0.5-ml microcentrifuge tube. Label it and set it aside for observation by microscopy later. Transfer all of the rest of the microsphere suspension to the 50-ml conical centrifuge tube.

   b. The freeze-dryer should have been turned on by your instructor ahead of time. Place the 50-ml conical centrifuge tube in liquid nitrogen to freeze the microsphere suspension. Tilt the tube to increase the surface area if possible. When the suspension is completely frozen, *loosely* cap the tube, and place it in a freeze-drying flask. Connect the flask to a freeze-dryer, and immediately turn on the vacuum.

5. *Observing the PLGA microspheres using an optical microscope*
   a. Clean and dry a glass slide and a coverslip with 70% ethanol. Pipette 20 μl of the microsphere suspension from Step 4a onto the slide and place a coverslip on top.
   b. Place the slide on the microscope stage. Use 200x or 400x magnification to observe the microspheres and take pictures. If the density is too high to see individual microspheres, then dilute the suspension ten times with de-ionized water and repeat the observation.
   c. Take pictures of the stage micrometer using the same magnifications as used in the previous step. By comparing pictures of the microspheres and the micrometer, the sizes of the microspheres can be estimated.
6. *Finishing up:* Wash all glassware and centrifuge tubes with dish detergent, rinse with de-ionized water, and hang them up to dry. Wash the magnetic stir bars and dry them on Kimwipe tissue. Properly dispose of the microscope slide and coverslip. Shut off the microscope and replace the dust cover.

Session 2. Evaluation of the Encapsulation Efficiency and the Drug Load

We now have PLGA microspheres with encapsulated BSA. Or do we? Has the BSA actually been encapsulated in the microspheres? If so, how much of the BSA is encapsulated? Two parameters, the *encapsulation efficiency* and the *drug load*, are quantitative answers to these questions. Specifically, the encapsulation efficiency refers to the amount of encapsulated protein (or drug) as a percentage of the total amount used for the fabrication process; thus, it reflects how successful the encapsulation is. The drug load refers to the amount of encapsulated protein as a percentage of the total weight of the microspheres (which includes the weight of the protein), and this parameter measures the amount of protein (or drug) that is available for controlled release.

Our goal for Session 2 is to measure the encapsulation efficiency and the drug load of our microspheres. We will use a small amount of microspheres from the total yield for these experiments, which is not unlike taking samples from a production line for quality control. The most crucial measurement is determining the amount of BSA encapsulated in the microspheres. Therefore the encapsulated protein needs to be released from the polymer matrix first, and this is accomplished by dissolving the microspheres in dichloromethane. An aqueous buffer is subsequently used to extract the protein because BSA is not soluble in organic solvent but is soluble in water (Figure 2.8). We can then measure the concentration of the extracted protein using *the Bradford method for protein concentration determination*. This method is based on the binding of a dye, Coomassie blue, to proteins. This interaction changes the absorption maximum of Coomassie blue to 595 nm, which manifests as a deep blue color. To measure protein concentration, a calibration curve is first established by measuring the absorption of Coomassie blue at 595 nm in the presence of known concentrations of BSA. Other protein samples are measured in the same way, and their concentrations can be calculated using the calibration curve (Figure 2.9).

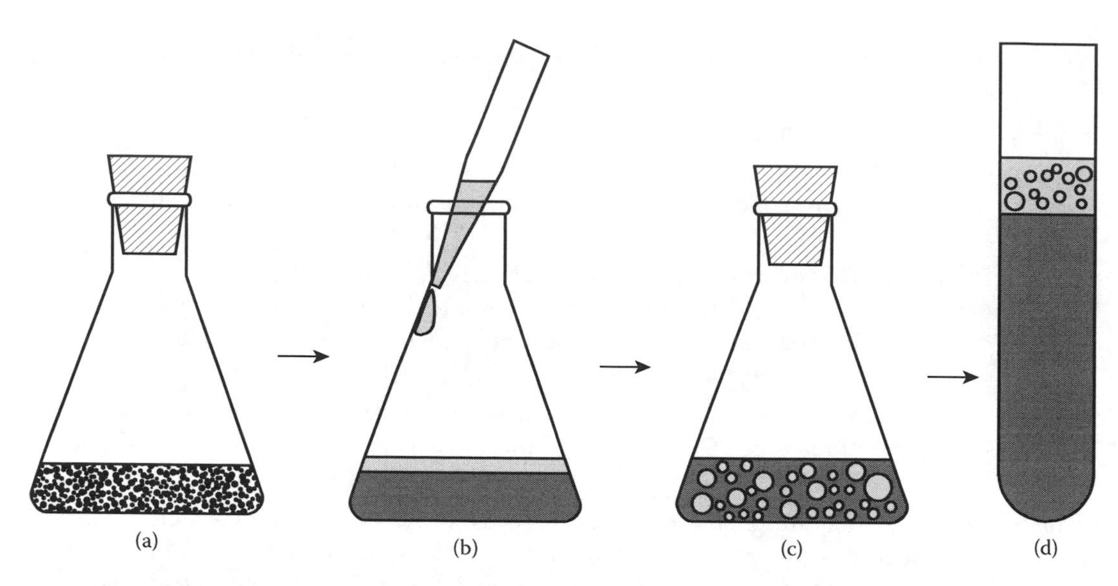

FIGURE 2.8
Extraction of encapsulated BSA from PLGA microspheres. (a) Measured amount of BSA-PLGA microspheres are added to dichloromethane and dissolved with shaking. (b) After the microspheres are completely dissolved, a measured volume of PBS is added. (c) With gentle swirling, PBS is broken into droplets and dispersed to extract the BSA. (d) After extraction is complete, the mixture is transferred to a test tube. A continuous aqueous phase will eventually form on top of the organic phase again. The concentration of this BSA solution is to be measured in order to determine the amount of BSA encapsulated in the PLGA microspheres.

We can then calculate the encapsulation efficiency (E.E.) using the following equation:

$$\text{E.E.} = \frac{\text{weight of encapsulated protein}}{\text{total weight of protein}} = \frac{C_{\text{extraction}} \times V_{\text{extraction}}}{C_{\text{total}} \times V_{\text{total}}} \times \frac{W_{\text{total}}}{W_{\text{extraction}}}$$

The variables in equation MI.1 are

$C_{\text{extraction}}$: the concentration of the extracted protein solution
$V_{\text{extraction}}$: the volume of aqueous buffer used for the extraction
$W_{\text{extraction}}$: the weight of the microspheres used for the extraction
C_{total}: the concentration of the BSA solution that was used in the encapsulation process
V_{total}: the volume of the BSA solution that was used in the encapsulation process
W_{total}: the weight of the total microsphere yield

And using the same variables, the drug load (D.L.) can be calculated using the following equation:

$$\text{D.L.} = \frac{\text{weight of encapsulated protein}}{\text{weight of microspheres}} = \frac{C_{\text{extraction}} \times V_{\text{extraction}}}{W_{\text{extraction}}} \qquad \text{(MI.2)}$$

Safety Note

Dichloromethane is again used in this session. Wear nitrile gloves, which are resistant to dichloromethane. Do ***not*** wear latex gloves. All procedures concerning dichloromethane should be carried out in a fume hood.

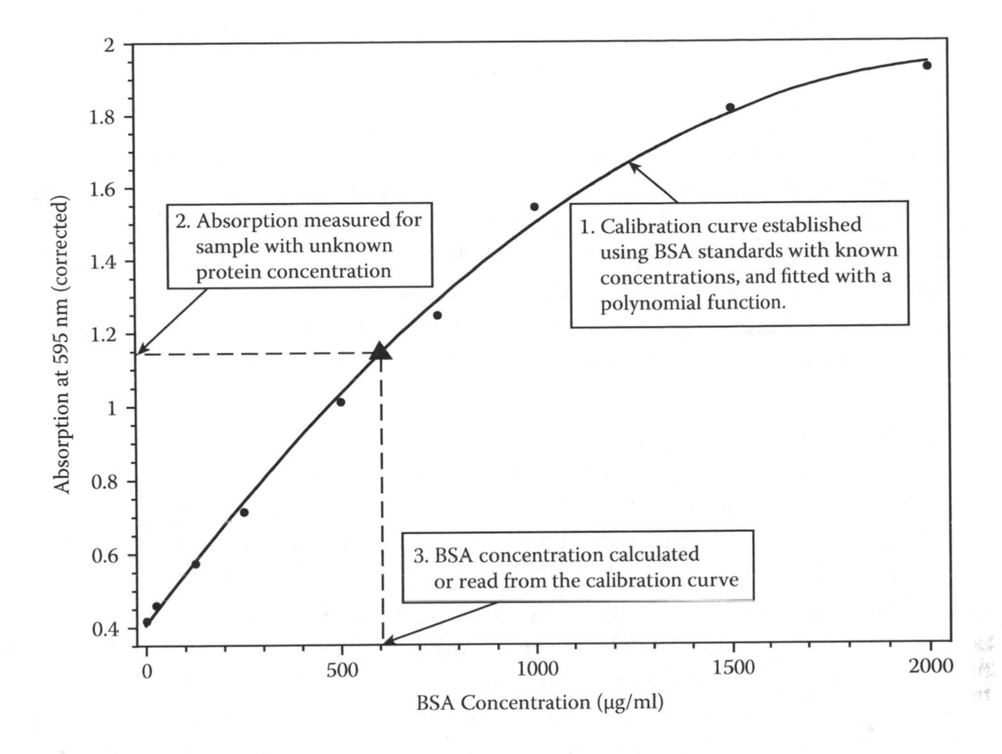

FIGURE 2.9
Using a calibration curve to measure protein concentrations. Step 1: A series of BSA standard solutions with precise concentrations are made through dilution from a BSA solution with a precisely measured concentration. The BSA standards interacted with the Bradford Coomassie reagent to produce a deep blue color with a maximum absorption at 595 nm. The absorbance for each standard is measured and plotted against the concentrations to produce the calibration curve, which is fitted with a polynomial function. Step 2: To measure a protein sample with an unknown concentration, allow the same volume of the protein sample to interact with the same volume of the Coomassie reagent as in Step 1, and then measure the absorbance at 595 nm. (The calibration and the measurement should be made at the same time.) Step 3: Once the absorbance is measured for the "unknown"protein sample, its concentration can be calculated using the fitted polynomial function or directly read off the curve.

Procedures

1. *Check-in*
   a. Samples and materials
      - Freeze-dried PLGA microspheres (from Session 1)
   b. Reagents
      - Dichloromethane
      - 50 mg/ml BSA (from Session 1, Part 1, Step 4)
      - BSA standard, 2000 μg/ml
      - PBS with 0.1% sodium azide
      - Bradford Coomassie reagent
   c. Special equipment and supplies
      - Glass serological pipettes (nondisposable)
      - Disposable glass test tubes (diameter ~1.3 cm, length ~10 cm)

- 25-ml Erlenmeyer flask with silicone stopper
- Disposable plastic cuvettes, 1.5-ml
- Spectrophotometer

2. *Measuring the total yield of the microspheres:* Zero an *analytical balance* with a weighing boat, and then transfer the PLGA microspheres from the 50-ml conical tube to the weighing boat. Record the weight of the microspheres. This is the total yield of the microspheres, or W_{total} in Equation MI.1.
3. *Taking a microsphere sample:* In a small weighing boat, weigh about 10 mg of the PLGA microspheres. Record the actual weight in full precision (0.1 mg). This will be the weight of the microspheres used for the extraction, or $W_{extraction}$ in Equations MI.1 and MI.2.
4. *Extracting the encapsulated BSA*
   a. Transfer the microsphere sample from Step 3 into the 25-ml Erlenmeyer flask. Use a *glass* serological pipette to add 5 ml of dichloromethane to the flask. Stopper the flask and swirl vigorously until the microspheres are dissolved (Figure 2.8a).
   b. Add 1000.0 µl of PBS with 0.1% sodium azide to the flask (Figure 2.8b). Swirl gently for 5 minutes to extract the BSA (Figure 2.8c). Avoid vigorous shaking in order to minimize foaming, which can lead to protein denaturation.
   c. Use a *glass* transfer (Pasteur) pipette to transfer the mixture into a disposable glass tube and seal the tube with Parafilm. Do not disturb the tube until a clear aqueous phase appears on top of the organic phase (Figure 2.8d).
5. *Diluting the 50 mg/ml BSA solution:* In a 1.5-ml microcentrifuge tube, add 980.0 µl of PBS, and then add 20.0 µl of the 50 mg/ml BSA solution from Session 1, Part 1, Step 4; be sure to pipette up and down to rinse the protein solution completely out from the pipette tip. Vortex the microcentrifuge tube to mix, and label it as "BSA 50xd C_{total}." *Note:* "50xd" stands for "50 times diluted"; the 50 mg/ml BSA solution needs to be diluted 50 times so that the diluted solution is within the range of the BSA standards (see Step 6). You *must* factor in this dilution when calculating C_{total}.
6. *Preparing protein standards*
   a. *Note:* For protein concentration determination using the Bradford method, it is critical to establish an accurate calibration curve, which in turn depends on the accuracy of the concentration of the standards. Since we will make our BSA standards by dilution of a 2000 µg/ml standard, make very sure that you are pipetting the correct volumes, since the errors will accumulate. Also, protein solutions tend to be viscous; to ensure full delivery of the correct volume, rinse the pipette tips by pipetting up and down several times after delivering the bulk of the BSA solution.
   b. Set nine 1.5-ml microcentrifuge tubes on a tube rack. Label the tubes on the lid with A, B, C, D, E, F, G, H, and I, and then label these tubes on the side with 2000 µg/ml, 1500 µg/ml, 1000 µg/ml, 750 µg/ml, 500 µg/ml, 250 µg/ml, 125 µg/ml, 25 µg/ml, and 0 µg/ml in the order of A to I.
   c. If the 2000 µg/ml BSA standard solution comes in an ampoule, carefully break the ampoule. (No special tool required.) Pipette ~250 µl of the solution into the microcentrifuge tube labeled as "A."

TABLE 2.1

Dilution Scheme for Making BSA Standards

Microcentrifuge Tube Name	Volume of PBS	Volume of BSA Solution	Final BSA Concentration
A	0 µl	250 µl from Ampoule	2000 µg/ml
B	25 µl	75 µl from tube A	1500 µg/ml
C	65 µl	65 µl from tube A	1000 µg/ml
D	35 µl	35 µl from tube B	750 µg/ml
E	65 µl	65 µl from tube C	500 µg/ml
F	65 µl	65 µl from tube E	250 µg/ml
G	65 µl	65 µl from tube F	125 µg/ml
H	80 µl	20 µl from tube G	25 µg/ml
I	100 µl	0 µl	0 µg/ml (blank)

TABLE 2.2

Scheme for Mixing BSA Samples with Bradford Reagent

Glass Test Tube Name	BSA Samples and Volume
1	0 µg/ml BSA standard (tube I), 50 µl
2	25 µg/ml BSA standard (tube H), 50 µl
3	125 µg/ml BSA standard (tube G), 50 µl
4	250 µg/ml BSA standard (tube F), 50 µl
5	500 µg/ml BSA standard (tube E), 50 µl
6	750 µg/ml BSA standard (tube D), 50 µl
7	1000 µg/ml BSA standard (tube C), 50 µl
8	1500 µg/ml BSA standard (tube B), 50 µl
9	2000 µg/ml BSA standard (tube A), 50 µl
50xd C_{total} #1	"BSA 50 xd C_{total}" (from Step 5), 50 µl
50xd C_{total} #2	"BSA 50 xd C_{total}" (from Step 5), 50 µl
$C_{extraction}$ #1	Supernatant from Step 4c, 50 µl
$C_{extraction}$ #2	Supernatant from Step 4c, 50 µl

d. Prepare BSA standards according to Table 2.1. Again, be sure to pipette up and down for full delivery of the solution, and vortex each tube (except tube A) for complete mixing before making the next one.

7. *Preparing samples with Bradford reagent*

a. Label nine *disposable* glass test tubes serially from 1 to 9 for the calibration samples. Label two more tubes as "50xd C_{total} #1" and "50xd C_{total} #2" for duplicate measurements of the sample prepared in Step 5. Finally, label two other tubes as "$C_{extraction}$ #1" and "$C_{extraction}$ #2" for duplicate measurements of the extracted BSA.

b. Add 1500 µl Bradford Coomassie reagent into each of the test tubes from Step 7a: First add 1000 µl into each tube, then adjust the micropipette and add 500 µl into each tube in a second round.

c. Add BSA samples into corresponding tubes containing Bradford reagent according to Table 2.2. Again, pipette up and down to rinse the protein solution completely out of the pipette tips. Mix the solutions with a vortex mixer.

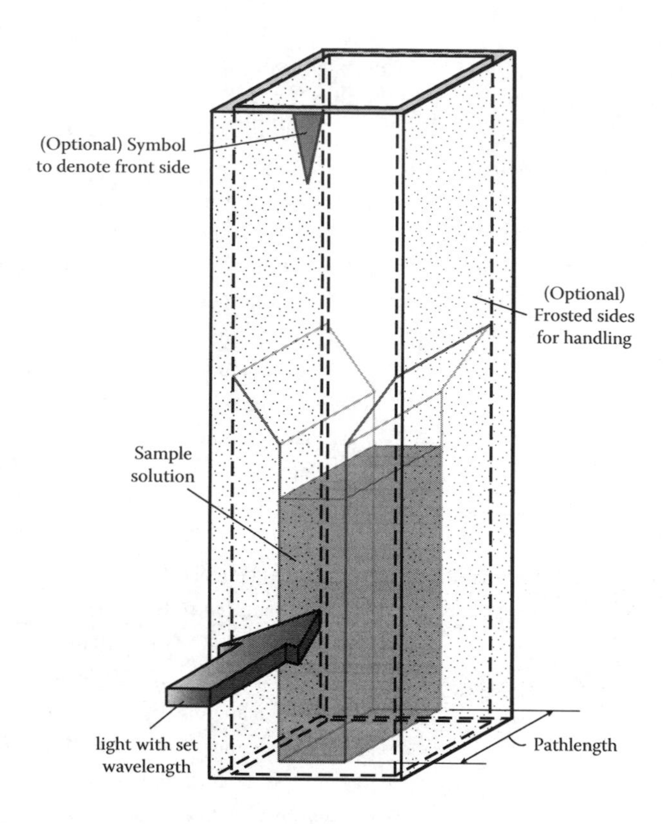

FIGURE 2.10
General features of a cuvette. A cuvette can be made of glass, quartz, plastic, or other specialized material (such as salt) and is used to hold samples for spectroscopic measurements. A light beam with a set wavelength passes through the sample at a defined path length, and the absorption can be measured by comparing the light intensity before and after passing through the sample. The cuvette depicted here is configured to reduce the amount of sample solution required.

d. Set a timer for 10 minutes. This allows time for the BSA-Coomassie complex to stabilize. (It is OK to leave the samples for longer than 10 minutes if, for example, the spectrophotometer in the next step is not available right away.)

8. *Calibration and measurement*

a. *Note 1:* A cuvette is a specialized container made of quartz, glass, clear plastic, etc. to hold liquid sample for spectroscopic measurement. The path length of a cuvette is the length that the UV-visible light from the spectrometer passes through, and the path length determines the absorption. A commonly used path length is 10 mm. (Some types of cuvettes offer two different path lengths depending on the orientation of the cuvette.) When handling cuvettes, do not touch the surfaces that will be in the light path. Some cuvettes have frosted sides for handling (Figure 2.10).

b. *Note 2:* Use the spectrophotometer according to its user's manual. Most models have a "blank" function, allowing you to set the baseline measurement to zero for convenience. Record the make and model of the spectrophotometer that you use in your lab notebook.

c. Pipette 2 ml of PBS into a disposable cuvette. Place the cuvette firmly into the cell holder, and make sure the long path (10 mm) of the cuvette is aligned with the light path. Wait for the absorbance reading to stabilize, and then blank the cuvette. Pipette the PBS out of the cuvette, but do not discard the cuvette (yet).

d. Vortex test tube "1" from Step 7 for at least 10 seconds to disperse any aggregate that might have formed, and then pipette the sample into the "blanked" cuvette. Place the cuvette in the spectrophotometer and wait for the absorbance reading to stabilize. Record the absorbance. When finished, discard the sample solution into a waste bottle and the cuvette into regular trash.

e. Repeat Steps 8c and 8d with the rest of the samples from Step 7. Use a new cuvette for each measurement.

f. Check and see whether the absorbance for the "50xd C_{total}" samples and the "$C_{extraction}$" samples is within the range measured for the BSA standards. If they are too high, make appropriate dilution and repeat the last steps.

9. *Finishing up:* Save your group's BSA standards and other BSA samples in a 4°C refrigerator. (You still need them for Session 4.) Wash and rinse the 25-ml Erlenmeyer flask and hang it up to dry. Wash the spatula and dry it on tissue paper. Make sure that the Bradford reagent samples are collected to the waste bottle and the used cuvettes are discarded in regular trash. Collect dichloromethane in a waste bottle. Discard the glass test tubes and the glass transfer pipettes in the broken glass disposal. Place the Bradford reagent and the BSA standard back to 4°C.

Data Processing

1. Data inventory: You should have the following data:
   - Weights: $W_{extraction}$ and W_{total}
   - Volume: $V_{extraction}$
   - Absorbance measurements: nine measurements for the calibration curve, two measurements each for C_{total} (50xd) and $C_{extraction}$
2. Plotting the calibration curve: Plot absorbance vs. BSA concentration, and fit the data points with a polynomial function (see Figure 2.9). Try functions of different orders to get the best fit. For example, Figure 2.9 is fitted with a second-order (quadratic) polynomial ($y = -3.23 \times 10^{-7}x^2 + 1.41 \times 10^{-3}x + 0.406$), and the example in Session 4 (see Figure 2.13) is fitted with a third-order (cubic) polynomial.
3. Obtain $C_{extraction}$ and 50xd C_{total} using the curve fit function. (You might need to use a math program for this task, or you can simply use a ruler to read the concentrations of the calibration curve.) Remember to convert 50xd C_{total} to C_{total}.
4. Calculate the E.E. and D.L. using Equations MI.1 and MI.2, and convert the results to percentages. Tabulate all measurements and results.

Session 3. Controlled Release of BSA from PLGA Microspheres

Many factors such as the molecular weight of the matrix polymer, the drug load, porosity, the size of the microspheres, etc. can affect the release of encapsulated agents from PLGA microspheres (Freiberg and Zhu 2004). In order to understand the release behavior of a given drug delivery device, a release profile needs to be established, and it can be *in vitro* or *in vivo* based. In typical *in vitro* study of drug release from microspheres, a given amount of the microspheres is suspended in buffer and shaken at 37°C, conditions that simulate the physiological environment of the human body. The buffer is sampled at regular time intervals to test for the amount of drug released. In *in vivo* studies, the microspheres are administered in animal subject (or in some cases, in human subjects), and tissue samples such as blood are taken at time intervals to test either directly for the released drug or indirectly for the molecule that the body produces as downstream response. An example of the latter is the antibody that is produced by the body as a response to a vaccine administered through controlled release (Katz et al. 2003). For PLGA microsphere systems, good correlation between *in vitro* and *in vivo* drug release has been observed (Abazinge et al. 2000).

In this session, we will set up an *in vitro* experiment to study the release of BSA from the PLGA microspheres that we have fabricated. By nature, controlled release studies take prolonged periods of time. Therefore, we will be collecting samples of released protein throughout this course: We will set up the *in vitro* controlled release by incubating the microspheres in PBS with gentle shaking. The released protein will be allowed to accumulate in the solution for a time interval, which is typically 1 week. The microsphere suspension is centrifuged each week, and the supernatant is collected. The microsphere sample is replenished with fresh PBS for the next sampling, and so on (Figure 2.11). This method of sampling simulates the constant dissipation of the released drug inside the body. We will set up duplicate samples for error evaluation of the data. The released BSA samples will be kept at 4°C until the end of the experiment, when we measure the concentrations of all the samples.

Procedures

1. *Check-in*
   a. Samples and materials
      - PLGA microspheres with encapsulated BSA (from Session 1)
   b. Reagents
      - PBS with 0.1% sodium azide
   c. Special equipment and supplies
      - Tabletop centrifuge with a rotor for microcentrifuge tubes
      - 37°C shaker incubator
2. In vitro *controlled release setup*
   a. *Note:* Sodium azide is toxic; it is used here as a biocide to prevent microbial growth.
   b. Weigh two batches of ~25 mg BSA-PLGA microspheres. Record the weights to 0.1 mg. Add each batch to a clean 1.5-ml microcentrifuge tube, and then add 1100 µl (100 µl + 1000 µl) PBS with 0.1% sodium azide to each tube. Pipette up and down to completely resuspend the microspheres. Label the tubes with your group's name and the date. These are your duplicate samples.
   c. Place the microcentrifuge tubes in a shaker incubator set to 37°C. Adjust the shaking speed so that the microspheres do not sediment.

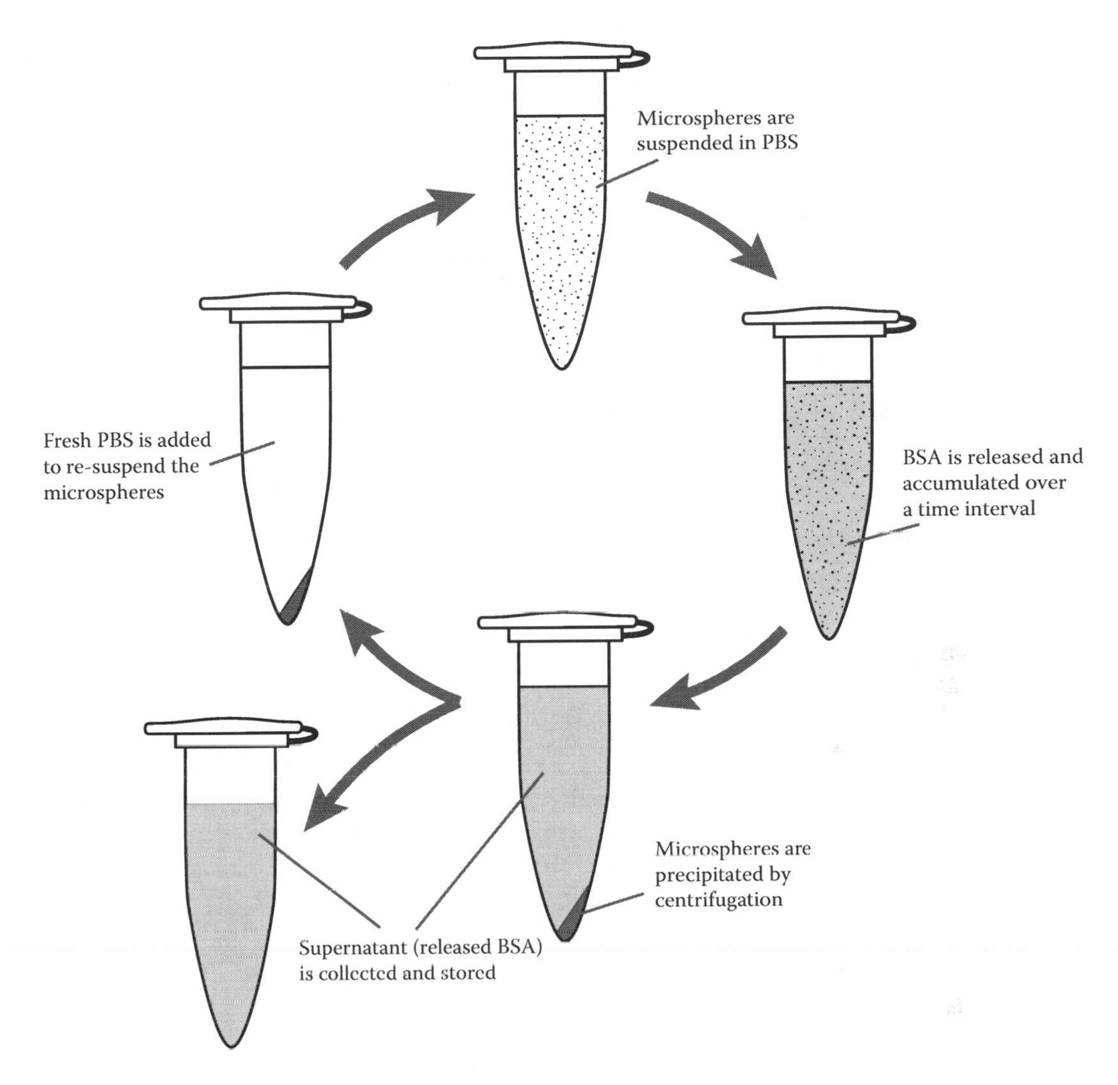

FIGURE 2.11
In vitro controlled release of BSA from PLGA microspheres.

3. *Taking controlled release samples*
   a. *Note:* Sampling at regular intervals for a prolonged period of time is the key to success for this experiment. Make sure to collect samples every week for the rest of the semester.
   b. Centrifuge the microsphere samples at 10,000 rpm for 1 minute. (Use longer time if the centrifugation speed is lower.) Your duplicate samples can balance each other for the centrifugation, but do eyeball the samples to make sure that they are about equal.
   c. Pipette 1000 µl of the supernatant from each tube and add to a 1.5-ml microcentrifuge tube. Do not disturb the microsphere pellets.
   d. Add 1000 µl of fresh PBS with sodium azide back to resuspend the microspheres. Place the microcentrifuge tubes back in the shaker incubator and continue shaking.

TABLE 2.3

An Example Schedule for Sample Collection for the Controlled Release Experiment

Week	Cumulative Days	Date of Sample Collection	Samples Collected By	Note
0	0	Jan. 31, 2008	W. Xian	O.K.
1	7	Feb. 7, 2008	A. Student	Someone turned off the shaker
...	...	...	...	...

e. Label the collected samples with your group's name and the date. Place the samples in a microcentrifuge tube rack labeled with your group's name; store the tube rack at 4°C.

f. *Note:* The above steps will be repeated every week for the rest of the semester. Each time point should have two duplicate samples. It is vital that you label the correct dates on the tubes and write them in your lab notebook. Use Table 2.3 to keep track of your group's sample collection; day 0 is the day you set up the controlled release experiment.

Session 4. Evaluation of Controlled Release Profile

We have been taking samples of BSA released from PLGA microspheres over the whole semester, and in this session, we will measure the BSA concentrations of these samples in order to obtain a controlled release profile. As in Session 2, we will use the Bradford method for protein concentration determination, but this time we will make the measurement with a different format. In Session 2, we used the "test tube protocol" in which samples were mixed (literally) in test tubes. That protocol was optimized for a protein concentration range of 125–1000 μg/ml. The test tube protocol offers better sensitivity and accuracy, but it takes time and effort since the samples have to be mixed and measured one by one. In this session, we will employ the "microplate protocol," in which a 96-well microplate will be used for sample preparation, and a microplate reader, which is essentially a spectrophotometer with specialized configuration, will be used to obtain the measurements. The microplate protocol is optimized for a protein concentration range of 1–100 μg/ml, and it offers better efficiency, which becomes important when there is a large number of samples to measure, as is the case here. Again, as in Session 2, we will first establish a calibration curve using BSA standards, and then determine the BSA concentrations of the collected samples using the calibration curve.

Ideally, the BSA release profile should resemble the solid line in Figure 2.12; in reality, however, release profiles oftentimes resemble the dotted line because of several factors. The first factor is the initial burst release (Figure 2.12, box 1), which refers to a relatively large amount of drug being quickly released when a drug delivery device is placed in contact with medium. This phenomenon has been observed not only for monolithic drug delivery devices such as the PLGA microspheres that we are working with but also for other drug delivery systems such as reservoir systems and hydrogel systems. For PLGA microspheres, one of the reasons for the initial burst release of protein is the formation of pores, cracks, and channels near the surface of the microspheres during solvent evaporation

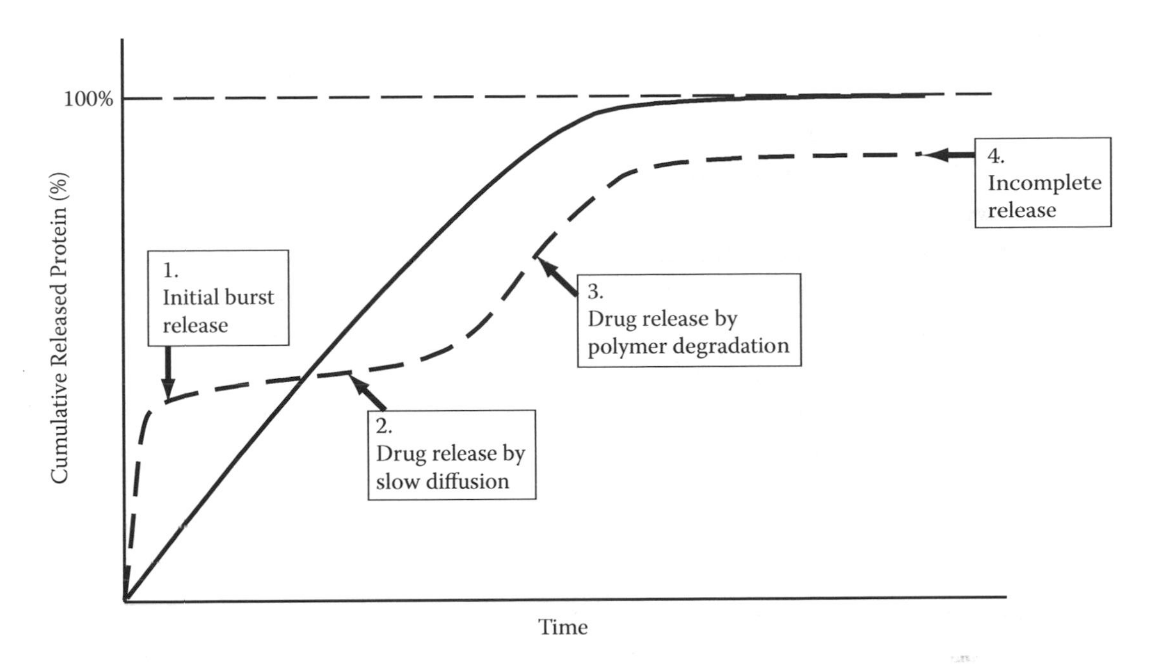

FIGURE 2.12
Ideal and non-ideal controlled release profiles. For an ideal controlled release (solid line), the drug is released at a constant rate until it has been completely released. But in reality, release profiles, especially for proteins, are usually non-ideal (dotted line); it may include an initial burst release, a slow release, a fast release, and then the release is sometimes incomplete due to deactivation or degradation of the protein.

(Huang and Brazel 2001). Burst release is often undesirable since it causes significant loss of the encapsulated drug from the drug delivery device, not to mention an unpredictable increase of drug concentration that can have harmful effect on the body. However, it can be of benefit in certain situations. For example, when interleukin-2 (IL-2) is delivered as an anticancer agent, its local concentration must be high enough to activate the immune system, and thus an initial burst of IL-2 followed by sustained delivery actually serves the therapy well (Thomas et al. 2004). Following an initial burst, in some cases there is a stage of slow and steady drug release (Figure 2.12, box 2), and such a stage is often observed for drugs with high molecular weights such as proteins. Since these molecules are too large to diffuse through the matrix, at this stage their release is due to the steady degradation of the surface of the matrix, or surface erosion. Meanwhile, a network of water-filled channels starts to form inside the microspheres, and when these channels are big enough, bulk erosion becomes prominent, which leads to the next stage of faster release (Figure 2.12, box 3). And finally, a factor that affects the release of proteins in particular is deactivation or degradation of the encapsulated protein (Figure 2.12, box 4), which often manifests as incomplete release of the protein. Deactivation of the encapsulated protein is potentially due to the microsphere fabrication process, where contact with organic solvent can cause protein denaturation. Another factor for protein deactivation or degradation may be the presence of lactic acid and glycolic acid, degradation products of PLGA, which can lower the local pH in the water-filled channels inside the microspheres to <3 (Fu et al. 2000; Shenderova et al. 1999). This acidic environment can cause deactivation or degradation of the encapsulated protein since many proteins are not stable in extreme pHs.

Procedures

1. *Check-in*
   a. Samples and materials
      – Collected samples of released BSA
   b. Reagents
      – BSA standards, 2000 μg/ml
      – PBS with 0.1% sodium azide
      – Bradford Coomassie reagent
   c. Special equipment and supplies
      – A 96-well plate with cover (do not touch the underside of the plate)
      – Multichannel micropipette and basins
      – Microplate reader
      – Timer
      – (Optional) Plate shaker
2. *Preparing BSA standards for the microplate protocol*
   a. *Note:* As in Session 2, BSA concentration accuracy is crucial; therefore, it is important to deliver accurate volumes when pipetting. This is especially important for the microplate protocol since the liquid volumes are small, and any loss of liquid will result in big errors later on.
   b. In a 1.5-ml microcentrifuge tube, add 150 μl of PBS, and then add 50 μl of 2000 μg/ml BSA standard. Remember to pipette up and down several times to rinse out all the protein in the pipette tip. Vortex the tube to mix thoroughly. Label the tube with "500 μg/ml," the BSA concentration after dilution.
   c. Prepare BSA standards according to Table 2.4. Be sure to pipette up and down for full delivery of the solution, and vortex each tube (except tube A) for complete mixing before making the next one. Label each tube with its BSA concentration.

TABLE 2.4
Dilution Scheme for BSA Standards

Tube	Volume of PBS	Volume BSA Solution	Final BSA Concentration
1	720 μl	180 μl of 500 μg/ml	100 μg/ml
2	500 μl	500 μl from tube 1	50 μg/ml
3	200 μl	200 μl from tube 2	25 μg/ml
4	480 μl	320 μl from tube 2	20 μg/ml
5	280 μl	120 μl from tube 2	15 μg/ml
6	400 μl	400 μl from tube 4	10 μg/ml
7	300 μl	300 μl from tube 6	5 μg/ml
8	200 μl	200 μl from tube 7	2.5 μg/ml
9	400 μl	0 μl	0 μg/ml (blank)

3. *Preparing the microplate:* Designate wells in the 96-well plate for calibration and measurements. You can mark on the cover of the microplate with a permanent marker, but be sure to draw a diagram in your notebook showing which wells are designated for which samples. Take the following into consideration when you "map out" the microplate:
   - 150 µl Bradford reagent and 150 µl protein solution (either from the standard solution or the controlled release samples) will be mixed in each well.
   - Duplicate samples need to be prepared for each point of the calibration curve.
   - The controlled release samples have been collected in duplicates; each sample needs will use one well.
4. *Preparing samples with Bradford reagent in the microplate*
   a. *Note:* A multichannel micropipette can be used for efficient liquid handling for a 96-well microplate. Disposable basin is used to hold liquid for multichannel pipetting. You might want to practice pipetting with the multichannel micropipette several times before you handle the samples.
   b. Add 150 µl of Bradford reagent to the designated wells.
   c. Add BSA samples into corresponding wells. Mix the solutions by pipetting up and down several times. Pipette slowly to avoid generating too many bubbles. Cover the plate when finished.
   d. Set a timer for 10 minutes. This allows time for the BSA-Coomassie complex to stabilize. After 10 minutes, check and make sure that no bubble is present in any of the wells. (Bubbles can reflect light, leading to erroneous absorption measurement.)
5. *Reading the microplate:* There are a variety of microplate readers available. Follow the user's manual for your microplate reader with these guidelines:
   a. Allow the microplate reader to "warm up," or to stabilize.
   b. Set the wavelength for absorbance measurement. The optimum wavelength is 595 nm. (If your model is not equipped with a 595 nm filter, choose the nearest one that is <595 nm, such as 575 nm, for example.)
   c. Set a mixing (or shaking) time of 10 seconds if the microplate reader has a built-in shaking mechanism; this is to disperse any aggregate that may have formed during incubation. Otherwise, shake the microplate using a *microplate* shaker. (Do not use common laboratory orbital shakers.)
   d. Take three readings of the microplate, i.e., run the microplate three times through the reader, and use the average of these three readings for the rest of the calculations.
6. *Finishing up:* Transfer the liquid from the microplate to the proper waste container and discard the plate in regular trash. Properly dispose of all BSA samples and discard the microcentrifuge tubes in regular trash.

Data Processing

1. Data inventory: You should have three sets of readings for the whole microplate. Take the average of the three readings. Now you have one averaged reading of the whole microplate, which should include: nine data points for the calibration curve

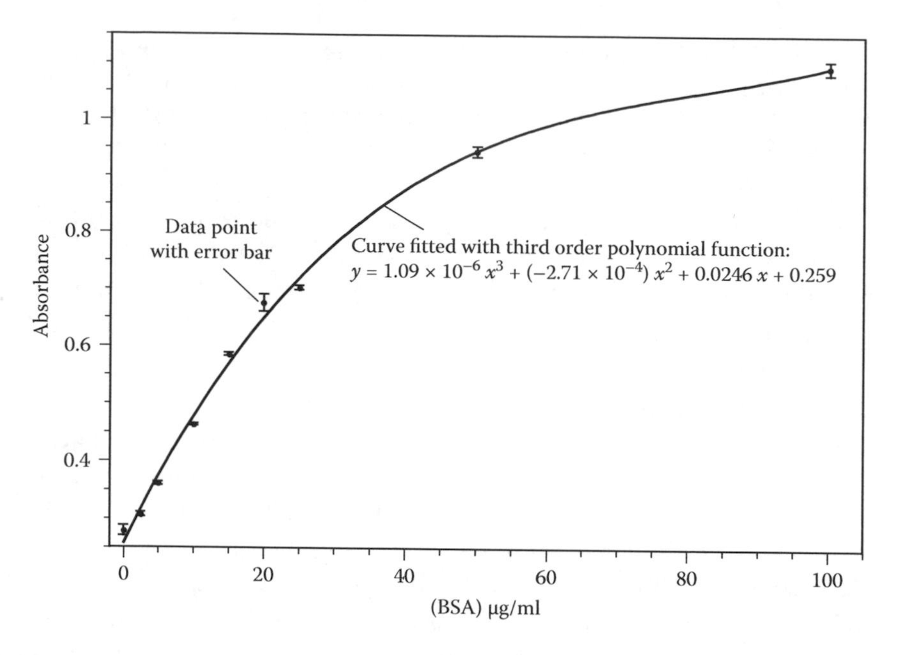

FIGURE 2.13
An example of a calibration curve for the microplate protocol. Note that (1) error bars are added since duplicate (or more) data are measured, and (2) the data points are fitted with a polynomial function (and a cubic polynomial works best for these particular data).

in duplicates, and n data points for the controlled release profile in duplicates, where n is the number of weeks that you have been collecting the released BSA.

2. Plotting the calibration curve: Calculate the average and the standard error of the two measurements for each point of the calibration curve, and plot absorbance vs. concentration using the averages; be sure to include error bars. Fit the data points with a best-fit polynomial function. An example is shown in Figure 2.13.
3. Plotting the controlled release curve: Follow the example in Table 2.5 to keep track of your data and results.
   a. After you have obtained the fit for the calibration curve, use the polynomial function to obtain the concentrations of your controlled release samples. Obtain the concentration for each sample individually.
   b. Use the concentrations of the duplicate samples to calculate the average and the standard error for each time point.
   c. Convert the concentration to weight with standard error for each time point. (Recall that 1000 μl of fresh buffer is used for each time interval except for the first one.)
   d. For each time point, calculate the cumulative weight of released BSA up to that time point.
   e. Convert the cumulative weight to cumulative percentage. Recall the amount of PLGA microspheres used for the controlled release and the drug load of the microspheres.
   f. Finally, plot the cumulative released percentage vs. the cumulative number of days. Remember to add error bars.

TABLE 2.5

(Example) Measuring the Amount of BSA Released from PLGA Microspheres Over Time

	Raw Absorption		Concentration[a] (μg/ml)		Cumulative Weight[b]		Cumulative Released[c]		Average Cumulative Released ± std. err.
Cum. Days	**#1**	**#2**	**#1**	**#2**	**#1**	**#2**	**#1**	**#2**	
0	—	—	—	—	0	0	0	0	0
7	1.104	1.128	106.60	108.00	117.26	118.8	4.62	4.77	4.72 ± 0.05
14	1.082	1.085	93.90	95.10	211.16	213.90	8.31	8.59	8.45 ± 0.14
...	...	...	...	...	...	...	...	...	...

[a] Concentrations are calculated using the polynomial function in Figure 2.13.

[b] The volume of PBS with sodium azide for the first time interval is 1100 μl (1.1 ml), and for the rest is 1000 μl.

[c] Assume that 25.4 mg of PLGA microspheres are used for sample #1, 24.9 mg are used for sample #2, and that the drug load is 10.0%. Thus the total amount of BSA available for release is 2540 μg for sample #1, and 2490 μg for sample #2.

Questions

1. Before the first-degree emulsion, the protein solution and the PLGA solution are both clear. What happens to the color (or transparency) when the two solutions are mixed by sonication, and why?
2. Why is 9% PVA solution used when making the second-degree emulsion? Can de-ionized water be used to substitute for the 9% PVA solution?
3. You are trying to harvest about 70 ml of PLGA microsphere suspension after a double emulsion process followed by solvent evaporation, but you only have 50-ml centrifuge tubes whose brim volume is 38 ml. You decide to distribute the suspension into two 50-ml centrifuge tubes, and since there is no more room in the tubes, you decide not to add de-ionized water to dilute the suspension. You then proceed to centrifuge the microspheres using the usual centrifugation speed and time. When finished, you find that the microspheres have not fully sedimented. Why is that?
4. Which of the following statements is (are) incorrect?
    a. In the first-degree emulsion, the aqueous phase, which contains protein, is in the form of tiny droplets.
    b. After the second-degree emulsion, the microspheres become hardened as dichloromethane evaporates.
    c. Polyvinyl alcohol (PVA) is used to stabilize the PLGA matrix so that the PLGA will not degrade during the subsequent fabrication processes.
5. You decide to use the double-emulsion method to encapsulate an antibiotic in PLGA microspheres. Suppose you start with 50 mg of PLGA dissolved in 5 ml of dichloromethane and 100 mg of antibiotic dissolved in 1 ml of PBS. After the double-emulsion process and subsequent freeze-drying, you obtain 160 mg of PLGA microspheres. The total weight of your starting materials is only 150 mg, so why is the yield greater than 100%?
6. In Session 2, why do we need to measure the concentration of the 50 mg/ml BSA solution using the Bradford assay? Do we not already know its exact concentration since we weighed the BSA to make this solution in Session 1?

7. Penicillin is still a widely used antibiotic for many common bacterial infections. Suppose we want to fabricate PLGA microspheres with encapsulated penicillin. In our experiment, 100.00 mg of penicillin and 50.00 mg of PLGA pellets are used, and 160.00 mg of penicillin-PLGA microspheres are obtained. We then use 10.00 mg of the penicillin-PLGA microspheres for analysis, and find 4.50 mg of encapsulated penicillin. What is the encapsulation efficiency and the drug load?
8. The following are three scenarios that can take place in lab:
   - Scenario 1: You are fabricating PLGA microspheres with encapsulated BSA, and for the step where you are supposed to add 5 ml of dichloromethane to dissolve PLGA microspheres, you accidentally add more than 5 ml of the solvent.
   - Scenario 2: Later, you are trying to determine the encapsulation efficiency and the drug load, and where you are supposed to add 1000 μl of PBS to extract the BSA, your co-worker accidentally adds more than that 1000 μl so that neither of you knows how much PBS has been added.
   - Scenario 3: You are trying to determine BSA concentrations. When setting up the color reaction, you are supposed to mix 1.5 ml of Bradford reagent and 50 μl of BSA standard solution for each sample, but because of a mix-up, you have added only 5 μl of the BSA solution instead.
   - Scenario 4: Similar to Scenario 3, but instead of adding 50 μl BSA standard solution, you have added 55 μl.

   For each scenario, does the mistake matter? If yes, what is the best way to remedy it?
9. The Bradford assay can be conducted using two different protocols: one is the so-called test tube protocol, which we used in Session 1, and the other is the microplate protocol, which we used in Session 4. What are the advantages and disadvantages of each one?
10. Insulin encapsulated in PLGA microspheres can potentially be used in therapy for patients with type I diabetes (Kang and Singh 2005). In one study, two different methods are used to encapsulate insulin, and the microspheres fabricated with these two methods give two different release profiles:
    - Method 1 involves using water-in-oil-in-water (*w/o/w*) double emulsions. Insulin solution is first emulsified with PLGA solution by sonication, and then PVA solution is added to create the second-degree emulsion by homogenization. In the release study, the initial burst release is 5%, and about 20% of the total insulin is steadily released over 2 months.
    - Method 2 uses solid-in-oil-in-water (*s/o/w*) emulsion. Insulin solid is first mixed with PLGA solution by sonication, and then PVA solution is added to create an emulsion through homogenization. In the release study, the initial burst release is 50%, and about 90% of the total insulin is steadily released over 2 months.

    The average size of the *w/o/w* microspheres is 55 μm, and that for *s/o/w* is 20 μm. From these results:

    a. Sketch the release profiles for these two studies. *Note:* Sketch the curves all in the same plot axes to show the differences qualitatively; it is also *crucial* to label the x and y axes of your plot.

b. Why is the release profile for the *s/o/w* microspheres different from the one for the *w/o/w* microspheres?

c. What kinds of problems would these two formulations post clinically?

d. Suggest a way to use these two kinds of microspheres together so that the released insulin is at clinically safe levels.

11. Experimental design: You need to design an injectable monolithic drug release system whose resorption time is about 12 months, and you have decided to use biodegradable microspheres. The drug to be released is a protein vaccine. Write a one-page proposal for a set of experiments that you need in order to fabricate and test this drug delivery system; include the following components in your design:

    - Rationale for using a microsphere delivery system
    - The material(s) of your choice and why
    - Outline of the fabrication process
    - Outline of how to test the release of the drug

    Note: When giving outlines of the experiments, there is no need to give operational details. However, you need to provide essential information about the experiments. For example, for the fabrication process, method of mixing, and whether it is sonication or homogenization, need to be specified; what solvent to use for dissolving the polymer of your choice is *very* important; for the release experiment, the duration of the test and the sampling frequency are also very important and need to be specified.

Appendix. Recipes and Sources for Equipment, Reagents, and Supplies*

Session 1. Encapsulation of BSA in PLGA microspheres

- **PLGA.** Lactel Absorbable Polymers, cat. no. B6010-2. *Selection criteria* (*Sel. crit.*): 50/50 poly(DL-lactide-*co*-glycolide) polymer with molecular weight (MW) ~70,000 Da.
- **BSA.** Fisher cat. no. BP1600-100. *Sel. crit.:* None.
- **PVA solution.** Dissolve PVA solid (Sigma-Aldrich, cat. no. 363170, *Sel. crit.:* MW 13,000–23,000) in de-ionized water with stirring. Filter with Büchner funnel and filter paper to remove insoluble matter.
- **20-ml glass vial.** Fisherbrand 20-ml borosilicate glass scintillation vials, Fisher cat. no. 03-337-15. (May be available in chemistry stockrooms.) *Sel. crit.:* Similar dimensions; solvent resistant.
- **Glass serological pipettes.** Kimax-51 reusable pipettes with standard opening, Fisher cat. no. 13-674-32H and 13-674-32J. *Sel. crit.:* None. Disposable glass pipettes are OK, too.

* *Disclaimer*: Commercial sources for reagents listed are used as examples only. The listing does not represent endorsement by the author. Similar or comparable reagents can be purchased from other commercial sources. See selection criteria.

- **Sonicator with microtip.** Fisher Scientific Model 100 Sonic Dismembrator, Fisher cat. no. 15-338-53. *Sel. crit.:* Power ~100 watts, and equipped with a fine tip probe (~1/8″).
- **Homogenizer.** Tissue-Tearor with 14 mm probe, BioSpec Products, Inc., cat. no. 985370-14. *Sel. crit.:* Similar output power (~140 watts) and probe dimension.
- **Centrifuge, rotor, and centrifuge tubes.** Eppendorf centrifuge 5810R with fixed-angle rotor (cat. no. F-34-6-38), capacity 6 × 85 ml. *Sel. crit.:* RCF >15,000 ×g; rotor capacity >50 ml per slot; refrigeration optional but preferred; select centrifuge tubes according to rotor.
- **Trip balance.** Ohaus Harvard trip balance, mfr. no. 1450-SD/EMD, Fisher cat. no. S40040.
- **Freeze-dryer:** Thermo Scientific ModulyoD freeze dryer with eight-port column manifold (cat. no. F05656000) and 1000-ml glass flasks (cat. no. F05657000). *Sel. crit.:* Bench-top models.
- **Optical microscope and image capture.** See BLS II appendix.
- **Microscope stage micrometer.** Meiji Techno America stage micrometer, mfr. no. MA285, Fisher cat. no. NC9167561. *Sel. crit.:* 0.01 mm division or less.

Session 2. Evaluation of the Encapsulation Efficiency and the Drug Load

- **Bradford Coomassie reagent.** Thermo Scientific Pierce Coomassie and Coomassie Plus—The Better Bradford Protein Assay Kit, mfr. no. 23236, Fisher cat. no. PI-23236. *Sel. crit.:* Protein concentration determination kit based on Bradford method; BSA standard solution included.
- **PBS with 0.1% sodium azide.** Make 10% sodium azide stock solution by dissolving 10.0 g of sodium azide (Sigma-Aldrich cat. no. S8032) in 100 ml of PBS. Add 1.0 ml of 10% sodium azide stock solution to each 100 ml of PBS.
- **Disposable glass test tubes.** Fisher cat. no. 14-958D. (May be available in chemistry stockrooms.) *Sel. crit.:* Similar dimensions.
- **Disposable plastic cuvettes.** Fisher cat. no. 14-385-942. *Sel. crit.:* Semi-micro (1.5-ml); light path 10 mm.
- **Spectrophotometer.** Thermo Scientific GENESYS 10 visible spectrophotometer, Fisher cat. no. 14-385-460. *Sel. crit.:* Range should be from 320 nm to 800 nm or higher.

Session 3. Controlled Release of BSA from PLGA Microspheres

- **Shaker incubator.** New Brunswick Scientific Model C76 water bath shaker, mfr. no. M1248 0002, Fisher cat. no. 14-278-180. *Sel. crit.:* Water bath preferred; temperature accuracy should be ±1°C or better.

Session 4. Evaluation of Controlled Release Profile

- **96-well plate.** *Sel. crit.:* Optically clear; not for tissue culture use; sterility not required.

- **Microplate reader.** Bio-Rad Model 680 microplate reader, cat. no. 168-1000EDU, with 575 nm filter, cat. no. 168-1045. *Sel. crit.:* Need a wavelength filter in the range of 570–600 nm.

References

Abazinge, M., T. Jackson, Q. Yang, and G. Owusu-Ababio, Comparison of *in vitro* and *in vivo* release characteristics of sustained release ofloxacin microspheres, *Drug Deliv.*, 7, 77–81, 2000.

Freiberg, S., and X.X. Zhu, Polymer microspheres for controlled drug release, *Int. J. Pharm.*, 282, 1–18, 2004.

Fu, K., D.W. Pack, A.M. Klibanov, and R. Langer, Visual evidence of acidic environment within degrading poly(lactic-*co*-glycolic acid) (PLGA) microspheres, *Pharm. Res.*, 17, 100–106, 2000.

Huang, X., and C.S. Brazel, On the importance and mechanisms of burst release in matrix-controlled drug delivery systems, *J. Control Release* 73, 121–136, 2001.

Kang, F., and J. Singh, Preparation, *in vitro* release, *in vivo* absorption and biocompatibility studies of insulin loaded microspheres in rabbits, *AAPS PharmSciTech.*, 6, E487–E494, 2005.

Katz, D.E., A.J. DeLorimier, M.K. Wolf, E.R. Hall, F.J. Cassels, J.E. van Hamont, R.L. Newcomer, M.A. Davachi, D.N. Taylor, and C.E. McQueen, Oral immunization of adult volunteers with microencapsulated enterotoxigenic escherichia coli (ETEC) cs6 antigen, *Vaccine*, 21, 341–346, 2003.

Putney, S.D., Encapsulation of proteins for improved delivery, *Curr. Opin. Chem. Biol.*, 2, 548–552, 1998.

Ratner, B.D., A.S. Hoffman, F.J. Schoen, and J.E. Lemons, *Biomaterials science: An introduction to materials in medicine,* 2nd ed., Elsevier Academic Press, Boston, 2004.

Shenderova, A., T.G. Burke, and S.P. Schwendeman, The acidic microclimate in poly(lactide-*co*-glycolide) microspheres stabilizes camptothecins, *Pharm. Res.*, 16, 241–248, 1999.

Thomas, T.T., D.S. Kohane, A. Wang, and R. Langer, Microparticulate formulations for the controlled release of interleukin-2, *J. Pharm. Sci.*, 93, 1100–1109, 2004.

3

Module II. Natural Biomaterials: Collagen and Chitosan

Collagen is the most abundant protein in humans and animals; it makes up more than 30% of the total protein mass by some estimates. Collagen molecules assemble to form protein fibers that are used as versatile building materials for the extracellular matrix and connective tissues such as skin, tendon, cartilage, and bone. In these tissues, collagen is organized in ways that give each tissue its unique mechanical properties. For example, in skin, collagen is organized into bundles of fibers that are interwoven with elastic fibers to give skin its mechanical toughness and resiliency. The bone is another tissue that shows the versatility of collagen as a building material: collagen forms a composite with mineral crystals, and while the inorganic phase contributes to the stiffness and strength of bone, the collagen phase adds flexibility and ductility (Lakes 1993).

By broad definition, collagen is the general name for a family of 28 proteins so far discovered that share similarities in amino acid composition, molecular structure, and function (Heino 2007). The basic collagen structural unit is tropocollagen (Figure 3.1), a long and rigid triple-stranded structure formed by three helically intertwining polypeptide chains. In these polypeptide chains, every third amino acid is a glycine, the smallest amino acid: Such arrangement of the amino acid sequence is the key for the triple-helix configuration of the tropocollagen. Proline, another structurally unusual amino acid, is also abundant in collagen. Hydroxylation of the proline and lysine residues allows extensive hydrogen-bond formation within and in between tropocollagen molecules, which is crucial to the structural integrity of the collagen. All collagen proteins share the function of serving as structural components of the extracellular matrix, but self-assemble in several different ways. Collagen that can form fibrils with a regular staggering pattern and show uniform cross-striation are called fibrillar collagen. The staggered arrays are reinforced by cross-linking among the tropocollagen units. The most abundant collagen, type I collagen, is a fibrillar collagen that is predominant in skin, bone, and tendon. Type I tropocollagen molecule is approximately 1.5 nm wide and 300 nm long—the aspect ratio is 200. Type I tropocollagen molecule is flexible with a reported persistence length of ~14.5 nm (Sun et al. 2002), but such flexibility is drastically reduced when tropocollagen molecules self-assemble to form fibrils.

Collagen as a biomaterial has many advantages and some disadvantages (Table 3.1). Currently many biomaterial applications of collagen are based on type I collagen, the most abundant and best-studied collagen, but biomaterials based on other collagen types are rapidly expanding. Below are examples of collagen-based applications approved by the Food and Drug Administration:

- Collagen or collagen composite sheets have been used as wound dressing for wounds, ulcers, burns, abrasions, etc. Commercial product of collagen wound dressing such as Fibracol® (Johnson & Johnson Corp., New Brunswick, New Jersey) has been shown to accelerate wound healing.

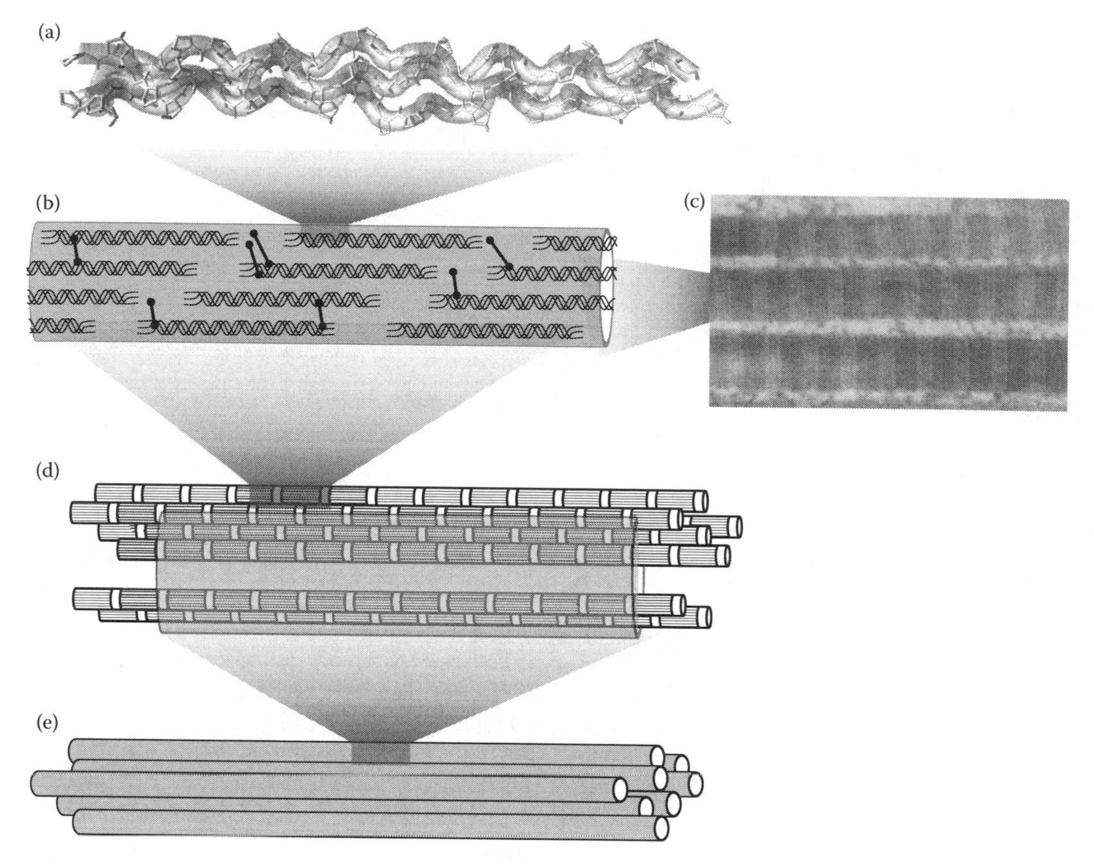

FIGURE 3.1
The hierarchical structures of collagen. (a) Tropocollagen is a basic structural unit of collagen. In this structural unit, three strands of polypeptides intertwine helically, and each strand itself is a helix. This structure is from Bella et al. 1994 (PDB ID: 1CAG). (b) A collagen fibril is a staggered array of tropocollagen molecules that interact with each other mainly through hydrogen bonds. The molecules can be cross-linked to further strengthen the fibril. Note that cross-linking frequently occurs at the two ends of the tropocollagen molecule. (c) Some types of collagen fibrils show a striation pattern in transmission electron microscopy (TEM) due to the periodic gaps in the staggered arrays of tropocollagen. Shown here is a TEM image of type I collagen from a mammalian lung tissue. (Image from Louisa Howard, http://remf.dartmouth.edu/images/mammalianLungTEM/source/9.html.) (d) Collagen fibrils assemble to form collagen fiber. (e) Collagen fibers can further assembly to form collagen fiber bundles.

- Collagen sponge has been used as hemostatic dressing because collagen can induce platelet aggregation that results in blood coagulation; in addition, it can also prevent postsurgical adhesion of tissues. An example is Superstat® (Interface Biomedical Laboratories Corp., Brooklyn, New York), a collagen derived topical hemostat.
- Collagen has long been used as injectable dermal filler for cosmetic correction or augmentation. In 2006, about 160,000 collagen injections have been performed in the United States at a cost of $250–$600 per treatment, out of which 53.5% used bovine-derived collagen, and 46.5% used human-derived collagen (http://www.collagen.org/cost.html). An example is Zyderm® (Inamed Corp., Santa Barbara, California), a commercial dermal filler based on bovine collagen.
- Since type I collagen consists of ~30% of the dry weight of bone, it is a natural choice to use as bone graft material. In one of these bone graft products, Infuse®

TABLE 3.1

Advantages and Disadvantages of Collagen as a Biomaterial (Lee et al. 2001)

Advantages
Available in abundance and easily purified from living organisms (constitutes more than 30% of vertebrate tissues)
Nonantigenic
Biodegradable and bioreabsorbable
Nontoxic and biocompatible
Synergic with bioactive components
Biological plastic due to high tensile strength and minimal expressibility
Hemostatic—promotes blood coagulation
Formulated in a number of different forms
Biodegradability can be regulated by cross-linking
Easily modifiable to produce materials as desired by utilizing its functional groups
Compatible with synthetic polymers
Disadvantages
High cost of pure type I collagen
Variability of isolated collagen (e.g., crosslink density, fiber size, trace impurities, etc.)
Hydrophilicity, which leads to swelling and more rapid release
Variability in enzymatic degradation rate as compared with hydrolytic degradation
Complex handling properties
Side effects, such as bovine spongeform encephalopathy (BSF) and mineralization

(Medtronic Sofamor Danek Inc., Minneapolis, Minnesota), a collagen sponge is not only used as scaffold for bone cells but also as a carrier for human bone morphogenetic protein-2, a signaling protein that promotes bone and cartilage formation.

Today collagen remains one of the most important biomaterials as more and more biomedical applications of collagen are being developed in areas such as drug delivery and tissue engineering. To explore new applications of this natural biomaterial, it is important to understand collagen as a protein and as a material. In this module, we will study type I collagen from calf skin through purification, biochemical characterization, and material fabrication.

Another natural biomaterial that we will study in this module is chitosan, which is a partially deacylated derivative of chitin, a structural polymer that is abundant in the exoskeletons of shellfish and insects (Figure 3.2). Due to its excellent biocompatibility and biodegradability, chitosan has been used in many biomedical applications, including (Dutta et al. 2002):

- Kidney dialysis membrane
- Wound dressings
- Resorbable sutures
- Contact lenses
- Drug delivery systems
- Space-filling implants

The activity and degradation of chitosan can be "tuned" by its degree of deacetylation, which can range from 50–90%, and its molecular weight, which has been observed to span from 300 kD to over 1000 kD. Chitosan and collagen can form composites

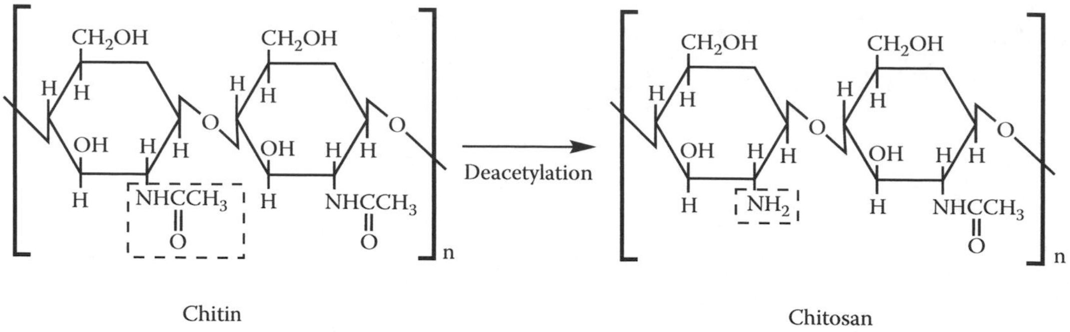

Chitin

Chitosan

Deacetylation

FIGURE 3.2
Structures of chitin and chitosan. Chitin can be converted to chitosan through deacetylation. As depicted, one of the two acetyl groups in the chitin unit is converted (dashed box), resulting in a chitosan unit with 50% deacetylation.

with improved mechanical and biological properties, and probably with economical benefits: chitosan is much less expensive than collagen due to its abundant availability. In our studies, we will use chitosan together with collagen to fabricate collagen/chitosan composite materials.

Telopeptides and Atelocollagen

The main source of antigenic activity for collagen is the short peptides, called telopeptides, at the two ends of the tropocollagen triplex helices. These telopeptides can be removed by pepsin, a protease that only removes the telopeptides but does not degrade the triple helices. Removal of the telopeptides can also reduce cross-linking since the linkages are usually made among the telopeptides. Pepsin is active in acidic solutions; therefore, the enzymatic treatment can be combined with the extraction process. Collagen with the telopeptides removed is called atelocollagen. Pepsin must be removed in the subsequent purification process. (Pepsin treatment will not be used in our experiments.)

Session 1. Extracting Acid-Soluble Type I Collagen from Bovine Calf Skin

Type I collagen is a fibrillar collagen, with tropocollagen units arranged in staggered arrays that in turn self-assemble to form fibrils. As mentioned earlier (see Figure 3.1), tropocollagen molecules can be cross-linked with covalent bonds; cross-linking is extensive in mature tissues, which enhances their tensile strength. Cross-linked collagen is not soluble unless enzymes or extreme chemical conditions are applied to break the cross-linking covalent bonds. On the other hand, in fetal or young tissues, a relatively high percentage of the collagen is not yet cross-linked, and individual tropocollagen molecules can be extracted with acid; the extracted collagen is call acid-soluble collagen. In acid extraction, the acidic condition changes the charges on the polypeptide chains and interferes with the hydrogen bonds that hold the tropocollagen molecules together; as a result tropocollagen molecules are released from the fibrils. However, the tropocollagen molecules must remain in an acidic environment to remain soluble; once the pH becomes neutral, they will self-assemble to form fibrils again. While this unique feature of type I collagen is sometimes undesirable—for example, the pH of injectable collagen must be acidic during storage, but it must be adjusted to neutral for the injection—it is especially helpful for cell encapsulation in tissue engineering, as we will see later in Module IV (Chapter 6). Acid-soluble collagen can be further purified. Pure acid-soluble collagen is especially suitable for applications such as injectable cosmetic implants, drug delivery, or tissue engineering. Collagen itself is considered weakly antigenic, and enzymatic treatment that removes the telopeptides can further reduce its antigenicity.

Calf skin is often used as a source material to obtain type I collagen. Other sources include tendons, eyes, or even placentas. In research laboratories, rat tail is a favorite source for type I collagen. Due to the threat of mammalian diseases such as bovine spongiform encephalopathy (mad cow disease) that can threaten humans, collagen from nonmammalian sources—deep-sea fish, for example—are also explored. In this session, we will use acetic acid, a relatively mild acid, to extract acid-soluble type I collagen from calf skin. Skin is one of the toughest tissues in animals and is very hard to process. Here we will use cryogenic pulverizer to freeze-fracture calf-skin into coarse powder to increase the extraction efficiency. Also, we must pay attention to the fact that collagen is a protein susceptible to denaturation. The tropocollagen molecules will especially become more labile after dissembly from fibrils. Interestingly, type I collagen begins to denature around 37°C, our own body temperature. Therefore, the ambient temperature should be kept cold or at least <25°C during the extraction and the subsequent processing to minimize denaturation.

Procedures

1. *Check-in*
   a. Samples and materials
      - Calf skin, trimmed to remove fat and muscle, and soaked in 0.01 *M* acetic acid to swell the tissue (Figure 3.3a).

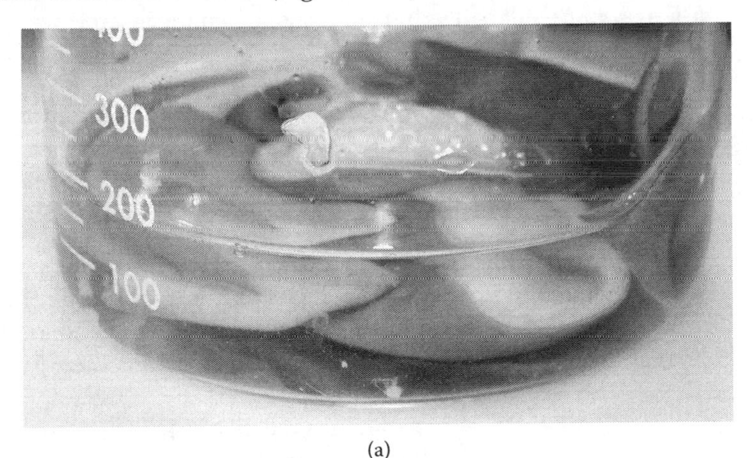

(a)

FIGURE 3.3
Extraction of type I collagen from calf skin in pictures. (a) Calf skin is shaved, cleaned, washed, and soaked in very mild acetic acid solution to swell the tissues.

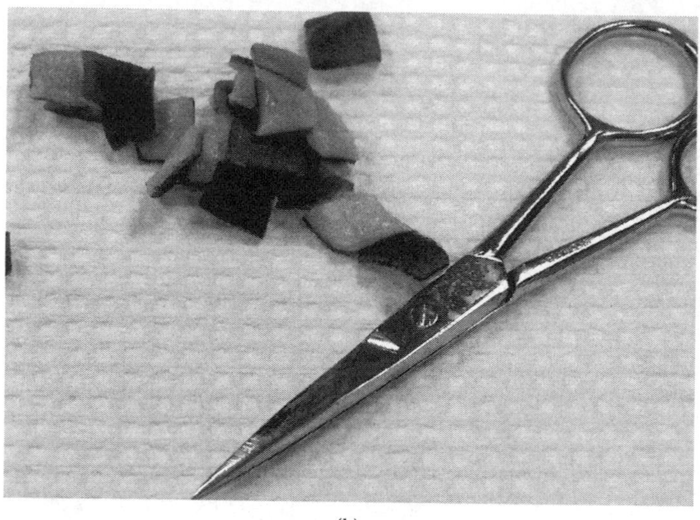

(b)

FIGURE 3.3 (continued)
(b) Calf skin is cut into 1 cm strips, and then 1 cm × 1 cm pieces.

b. Reagents
   - 0.5 *M* acetic acid
c. Special equipment and supplies
   - Two Dewar thermoses with liquid nitrogen
   - Scissors or scalpels
   - Tweezers
   - Cryogenic pulverizer
   - Erlenmeyer flasks, 250 ml

2. *Preparing calf-skin sample:* Cut calf skin into ~1cm × 1cm pieces and weigh about 5 g of the skin pieces (Figure 3.3b). Record the weight to 0.1 g. ***Caution:*** Handle sharp tools with care.
3. *Pulverizing calf skin*
   a. ***Caution:*** Handle liquid nitrogen carefully. Freeze the calf-skin pieces by adding them *one by one* into the liquid nitrogen to make sure that the pieces are not clumped together.
   b. Prepare the cryogenic pulverizer according to the user's manual. If using a hand-held pulverizer (BioPulverizer™, for example), submerge the mortar of the pulverizer in liquid nitrogen until the liquid nitrogen is no longer boiling. To avoid breaking it, make sure that the mortar is lowered and not dropped into the Dewar.
   c. For BioPulverizer or similar pulverizers: Make sure that the both the calf-skin pieces and the pulverizer mortar are both completely chilled. Remove the mortar with liquid nitrogen inside and place it on the lab bench. Chill the pestle using the liquid nitrogen inside the mortar until the liquid nitrogen boils off. Transfer one or two calf-skin pieces to the mortar, insert the pestle in the mortar, and deliver sharp blows by squeezing the spring-loaded handle.

(c)

(d)

(e)

FIGURE 3.3 (continued)
(c) Calf-skin pieces are chilled in liquid nitrogen and pulverized into coarse powder using a pulverizer. (d) Calf-skin powder is collected in a flask before thawing. (e) Calf-skin powder is suspended in acetic acid solution with stirring to extract type I collagen.

Rotate the pestle slightly between blows until the calf skin is shattered into coarse powder (Figure 3.3c). Quickly transfer the calf-skin powder to a 250-ml Erlenmeyer flask (before it thaws and becomes wet and sticky) (Figure 3.3d). Repeat until all the calf-skin pieces have been shattered. Add liquid nitrogen to the mortar in between the batches to keep it chilled.

4. *Extraction*
   a. Using a graduated cylinder, add 160 ml of 0.5 *M* acetic acid to the 250-ml flask. Also add a magnetic stir bar. Cover the mouth of the flask with Parafilm and label the flask with your group's name and the date.

   b. Place the flask on a stirring plate and adjust the stirring to a moderate speed (no visible whirlpool) (Figure 3.3e).
   c. The calf skin should be extracted for 48 hours or longer. If possible, keep it at 10°C or below; if not, make sure that the room temperature is <25°C.
5. *Finishing up:* Wipe away any residual skin tissue from the lab bench and the tools. (Lightly soiled paper towels can usually be discarded in regular trash, but make sure to follow your institution's regulations.) Wipe down the lab bench with 5% bleach or 70% ethanol. Wash the scissors or scalpel with dish detergent, rinse clean with de-ionized water, and leave on a paper towel to dry. Wash and rinse the graduated cylinder and place it on a drying rack.

Session 2. Purification of Extracted Collagen

We have now extracted acid-soluble type I collagen. But how good is this collagen? That is, is it biologically active, and is it pure enough? We will not be able to answer the first question yet; for the second question, one needs only to take a look at the extraction and will probably answer no—you can see that there are calf-skin residues that need to be removed, and there are noncollagen proteins that you cannot see but need to be removed as well. Our goal for this session is to remove the contaminating proteins released from the skin tissues along with type I collagen as outlined in Figure 3.4. We will use a method called salt precipitation to purify the extracted collagen, which is then dialyzed to remove salt and small molecules, and finally freeze-dried for long-term storage. The end product will be purified Type I collagen that retains full biological activity.

The salt-precipitation process that we will be using is also known as "salting-out" (Figure 3.5). In a protein solution with low ionic strength, the ionic or polar amino acids on the surface of the protein are solvated with water molecules, and the presence of ions helps to stabilize the protein by screening its charges. This is called "salting-in" (Figure 3.5a). As the salt concentration increases, the salt ions use up more and more water molecules for hydration, leaving less and less for hydration of the protein, thus effectively reducing the solubility of the protein. At sufficiently high salt concentrations, the solubility of the protein is reduced to below its concentration in the solution, and as a result, the protein precipitates out of the solution (Figure 3.5b). Meanwhile, the contaminating proteins, which presumably are present in smaller amounts, remain in the solution. Through this process, the target protein can be purified away from the contaminating proteins through precipitation. Of course, this method works only when the target protein is present in much higher concentration than the contaminating proteins. It is important to note that salt precipitation usually does not cause denaturation of the protein; therefore it is a reversible process: once the salt concentration is reduced, the precipitated protein will become soluble again. This salt-precipitation process can be used repeatedly to obtain higher purity for the target protein (although at a cost of lower yield).

The other method that we will use in this session is dialysis, a method that is also commonly used in protein biochemistry. In dialysis (Figure 3.6), a protein solution is placed inside a semipermeable membrane and is allowed to equilibrate with another solution, the dialysis buffer. The pores in the membrane will allow molecules smaller than the pores to pass, but larger molecules are retained inside the membrane. Dialysis is often used to remove small molecules such as salt from samples; it can also be used to change

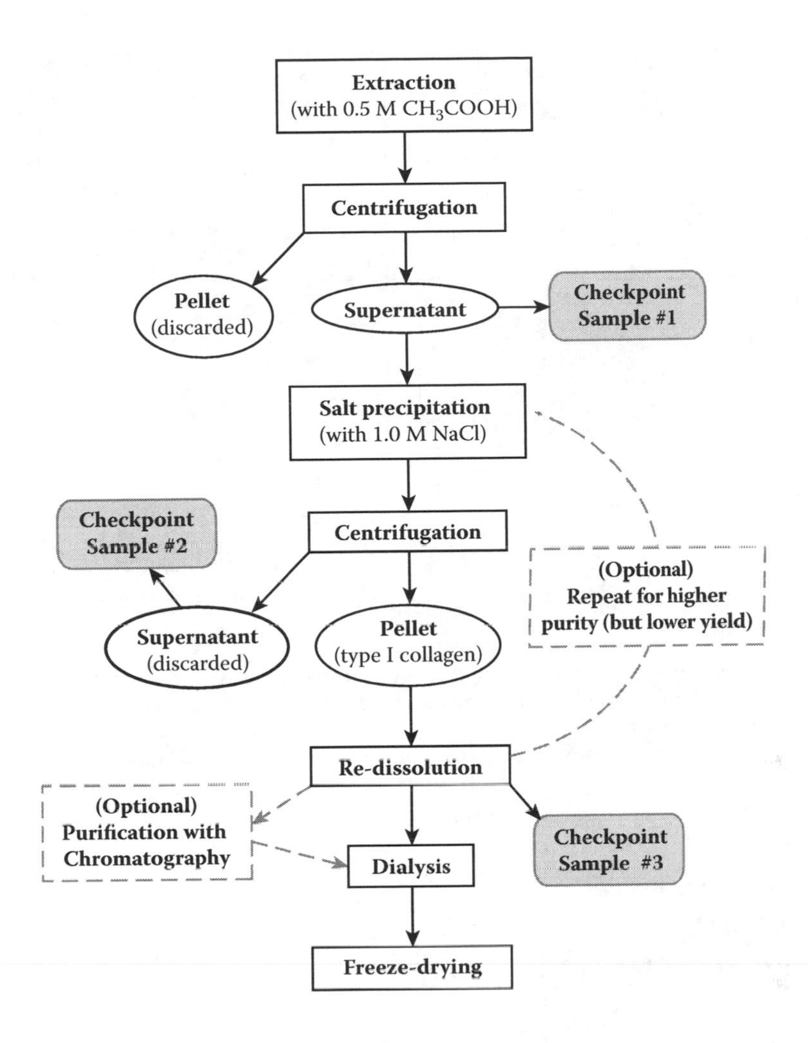

FIGURE 3.4
Flow chart of the procedures for collagen extraction and purification. Samples (in ovals) are taken at different steps (in rectangles) as checkpoint samples (rounded rectangles) and will be analyzed later to evaluate the efficiency of the process. Optionally, the salt-precipitation can be repeated to improve the purity of collagen, and the collagen can be further purified using other methods such as chromatography after salt precipitation.

the buffer composition of a protein solution to that of the dialysis buffer. Artificial kidney dialysis, the lifesaving clinical treatment for patients with kidney deficiency, operates on the same principle.

Note: Throughout the experiment, you will preserve a sample at each stage as a *"checkpoint" sample*. These samples will be analyzed later to check the effectiveness of each step in the purification process.

Procedures

Part 1 (Day 1). Salt Precipitation of Collagen

1. *Check-in*
   a. Samples and materials
      – Collagen extraction prepared in Session 1.

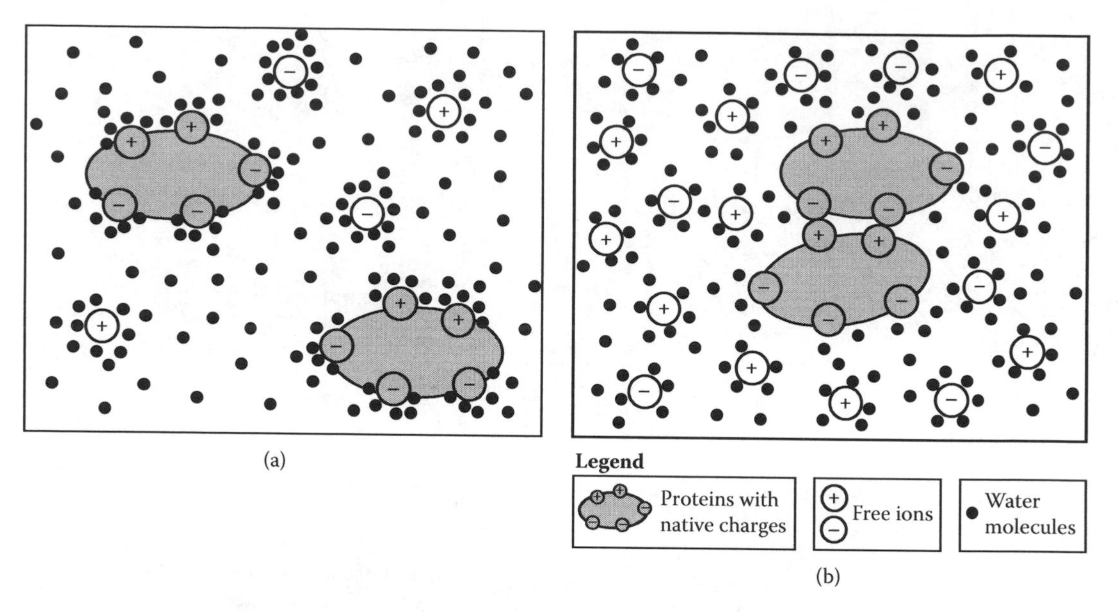

FIGURE 3.5
Salt precipitation of protein. (a) At low salt concentration, both the salt ions and the proteins are hydrated with water molecules, and the free ions help to stabilize the protein. (b) At very high salt concentration, the salt ions recruit large amounts of water molecules to their hydration shells, leaving fewer water molecules available to hydrate the protein molecules. The protein molecules aggregate and precipitate out of the solution as a result.

b. Reagents
   - NaCl solid
   - De-ionized water
   - 1.0 *M* HCl
   - 0.010 *M* HCl
   - 0.5 *M* acetic acid

c. Special equipment and supplies
   - Erlenmeyer flask, 250 ml (already used for extraction)
   - Beakers, 1000 ml, 250 ml, and 100 ml
   - Dialysis tubing and clamps
   - Round-bottom centrifuge tubes, 85 ml (actual brim volume 65 ml)
   - Centrifuge with fixed-angle rotor
   - Trip balance
   - Waste container for animal tissue

2. *Observation:* Your group's batch of pulverized calf skin should have been extracted in acetic acid for >48 hours at a temperature <25°C with moderate stirring speed at this point. Observe the extraction solution and note its color and viscosity.

3. *Removing insoluble matter*

   a. Distribute the extraction mixture to three 85-ml round-bottom centrifuge tubes. Rinse the flask twice with 5 ml 0.5 *M* acetic acid each and add the rinse to the centrifuge tubes as well. Balance one centrifuge tube against water or

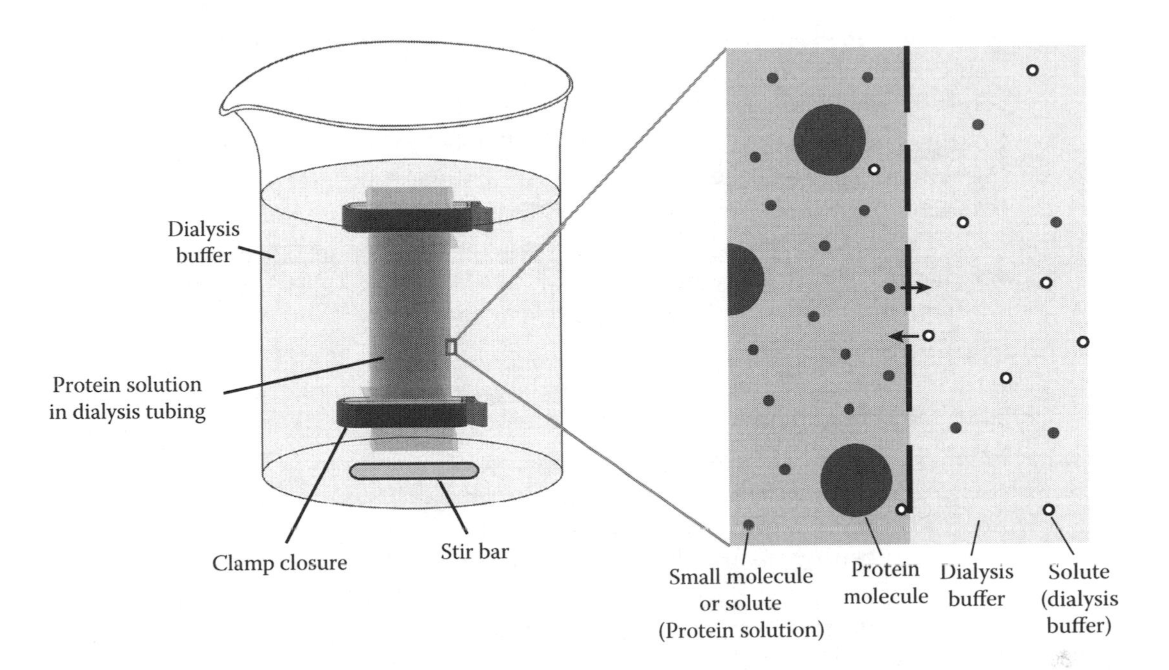

FIGURE 3.6
Dialysis in practice and in principle. In practice, a protein solution is sealed in a semipermeable membrane and placed in a large volume of dialysis buffer with stirring. The dialysis is complete when the exchange between the solutions inside and outside the membrane reaches equilibrium. In principle, small molecules can pass through the pores in the semipermeable membrane, while large molecules are retained. At equilibrium, the small-molecule compositions outside and inside the membrane become the same; thus, the small-molecule components of the dialyzed solution are exchanged for those of the dialysis buffer. The average pore size of the semipermeable membrane is characterized by its molecular weight cutoff (MWCO), which is defined as the molecular weight of a globular molecule that is 90% retained inside the membrane.

another group's tube (make sure that the caps are included), and then balance the other two tubes by transferring the liquid between them until balanced.

b. Centrifuge for 15 minutes at 10,000 rpm. Meanwhile, wash the used stir bar and flask with dish detergent and rinse with de-ionized water. Weigh the cleaned flask to 1 g accuracy.

c. After the centrifugation is finished, transfer the supernatant to the cleaned flask and weigh the flask again. Estimate the volume of supernatant by assuming that its density is close to that of water, 1.0 g/ml. (Set the centrifuge tubes aside for now; you will have time to clean them later after Step 5c.)

4. *Taking the first checkpoint sample:* Pipette 100 µl of the supernatant from Step 3c to a 1.5-ml microcentrifuge tube and mark it as "CP#1," along with your group's name and the date.

5. *Salt-precipitating collagen*

   a. Use the supernatant volume measured in Step 3c to calculate the weight of NaCl solid that is needed for a final concentration of 1.0 *M* in the solution. (The molecular weight of NaCl is 58.5.) Weigh the calculated amount of NaCl solid.

b. Add the cleaned stir bar back to the 250-ml Erlenmeyer flask and start stirring. Using a spatula, slowly add the NaCl solid into the collagen solution. Observe carefully what happens inside the collagen solution.

c. Continue to stir for 15 minutes after all the NaCl solid has been added. (Meanwhile, discard the pellets in the centrifuge tubes to the animal tissue waste container and wash the tubes. Afterwards, work on Steps 10a and b.)

6. *Harvesting the salt-precipitated collagen:* Transfer the collagen mixture from Step 5c to the cleaned 85-ml centrifuge tubes and balance the tubes. Centrifuge for 15 minutes at 10,000 rpm. (Meanwhile, clean the stir bar and the Erlenmeyer flask again.)
7. *Taking the second checkpoint sample:* After the centrifugation is complete, pipette 100 µl of the supernatant to a 1.5 ml microcentrifuge tube; mark it as "CP#2" and label it properly.
8. *Redissolving the salt-precipitated collagen:* Remove and discard the supernatant from Step 6. Add 5 ml of 0.010 *M* HCl solution to each tube and dissolve the pellet by stirring and pipetting up and down with a transfer pipette.
9. *Taking the third checkpoint sample:* When the collagen is redissolved, pipette 50.0 µl of the solution to a 1.5-ml microcentrifuge tube and mark it as "CP#3;" label the tube properly.
10. *Dialysis*

a. Record the molecular weight cut-off (MWCO), the vol./cm ratio (volume-to-length ratio), and the material of the dialysis tubing provided, and calculate the appropriate length for your sample. *Note:* The MWCO and the vol./cm ratio are two important parameters for selecting the right dialysis tubing. For most globular proteins, the MWCO should be half of the actual molecular weight of the molecule to be dialyzed. On the other hand, tropocollagen molecules happen to be rod-shaped, thus a much smaller MWCO is used here. The vol./cm ratio needs to be considered so that the right length for the dialysis tubing is used for a given volume. For example, if 25 ml of protein solution is to be dialyzed, and the vol./cm ratio of the dialysis tubing is 5.1 ml/cm, then the minimum length for the dialysis tubing should be 25/5.1 = 4.9 cm, but we must add at least 2 cm to each end of the dialysis tubing for clamping, thus the length of the dialysis tubing should be about 9 cm; to be safe, add another 10% to the minimum length; therefore we will measure and cut ~10 cm of the dialysis tubing.

b. Cut a piece of dialysis tubing with appropriate length and soak it in de-ionized water for at least 15 minutes.

c. In a 1000-ml beaker, add a stir bar, 990 ml of de-ionized water, and 10 ml of 1.0 *M* HCl (to make a 0.010 *M* HCl solution)—this is the dialysis buffer.

d. Once the dialysis tubing is fully hydrated, gently rub open one end of the tubing. Rinse both the outside and inside with de-ionized water. Fold over about 2 cm (or half an inch) of one end of the tubing and close it with a clamp. Make sure that the clamp is secure. Add the redissolved collagen solution into the tubing. Rinse each centrifuge tube with ~0.5 ml of 0.010 *M* HCl solution (from Step 10c) and add to the dialysis tubing as well. Once the tubing is filled, fold over 2 cm of the open end and close with another clamp.

e. Place the dialysis tubing in the 0.010 *M* HCl solution and cover the beaker with clear plastic wrap (such as Saran Wrap). Label the beaker with your group's name and the date. Stir the dialysis buffer at a medium speed on a magnetic stirring plate. Keep the temperature below 10° if possible; if not, then make sure that it is not >25°C.

f. Your instructor will change the dialysis buffer with fresh 0.010 *M* HCl solution twice during the next 2 days.

11. *Finishing up:* Place your checkpoint samples in 4°C. Clean the 250-ml flask and the centrifuge tubes again.

Part 2 (Day 2). Freeze-Drying Collagen for Storage

1. *Check-in*
   a. Samples and materials
      - Dialyzed collagen solution from Part 1.
   b. Special equipment and supplies
      - Liquid nitrogen
      - 30-ml round-bottom centrifuge tubes
      - 50-ml conical centrifuge tube
      - Centrifuge and trip balance
      - Freeze-dryer
2. *Removing insoluble materials (if any):* Your collagen should have been dialyzed three times in 1.0 L 0.010 *M* HCl solution. Transfer the collagen solution from the dialysis tubing to a 30-ml centrifuge tube. Balance it with another tube filled with water. Centrifuge for 10 minutes at 10,000 rpm.
3. *Freeze-drying the collagen*
   a. After the centrifugation is finished, transfer the supernatant to a 50-ml conical tube. Mark the conical tube with your group's name and the date.
   b. Freeze the collagen in liquid nitrogen and place it in the freeze-dryer. Make sure that the cap of the 50-ml tube is loose.
4. *Finishing up:* Rinse the used dialysis tubing and discard it in the regular trash. Clean the 1000-ml beaker and the centrifuge tubes.

Session 3. Analysis of Collagen Extraction and Purification by Electrophoresis

In previous work, we aimed to remove contaminating proteins from the collagen. But to what degree have we succeeded? To find out, we need to analyze the protein composition of the samples taken at various stages during the extraction and purification, i.e., the "checkpoint" samples, and we will use sodium dodecyl sulfate (SDS) polyacrylamide gel electrophoresis (PAGE), a method routinely used in biochemistry, to "see" the composition of the protein samples. This method is based on the following principles:

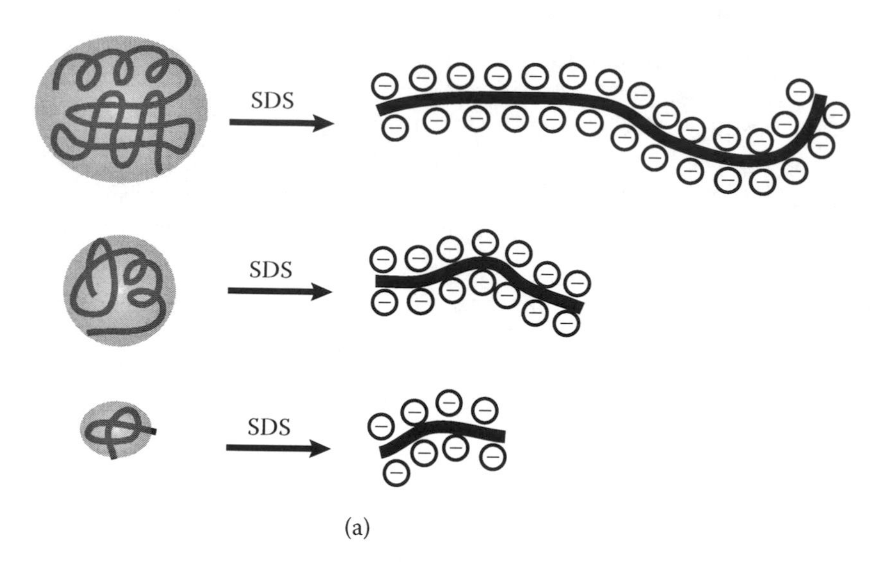

FIGURE 3.7
Principle of SDS-PAGE. (a) The three-dimensional globular structures of three different proteins are unraveled by treatment with SDS, and the proteins become flexible chains with negative charges. While the polypeptide chain lengths depend on the molecular weights of the proteins, the charge density is uniform for all three chains.

1. Simply speaking, electrophoresis is the movement of charged molecules in fluid toward opposite charges in the presence of an electric field.
2. To impart charges on protein molecules, SDS, a charged amphiphilic molecule, and a detergent, is used to destruct the three-dimensional configuration of the polypeptide chains by binding to them, and "coating" them with uniform charges (Figure 3.7a).
3. If nothing else stands in the way, molecules—large or small—will move with the same speed in an electric field if they bear the same charge-to-length ratio; therefore, a porous matrix is used to differentiate the molecules: small molecules will move through the pores easily, while the large ones will move more slowly because they are retained longer; thus, the relative mobility of a given molecule will depend on its molecular weight (Figure 3.7b). In SDS-PAGE, the polyacrylamide gel serves as the porous matrix.

When we run SDS-PAGE with a protein mixture, we can obtain information on 1) the molecular weights of the components, and 2) the relative amount of each component.

To analyze collagen as a protein, we need to first understand its protein characteristics. The acid-soluble type I collagen that we have purified consists mainly of tropocollagen. Of the three polypeptide chains in a tropocollagen unit, two are identical and are called the α1 chains, and the third one is called the α2 chain. Both the α1 and α2 chains have the same basic pattern for collagen amino acid sequence $(G\text{-}X\text{-}Y)_n$ in which G is glycine, and X and Y are other amino acids, but with different amino acid compositions. In SDS-PAGE, after the tropocollagen is denatured, the α1 chains and the α2 chain will be resolved as two bands on the gel due to the difference in electromobility between these two types of chains.

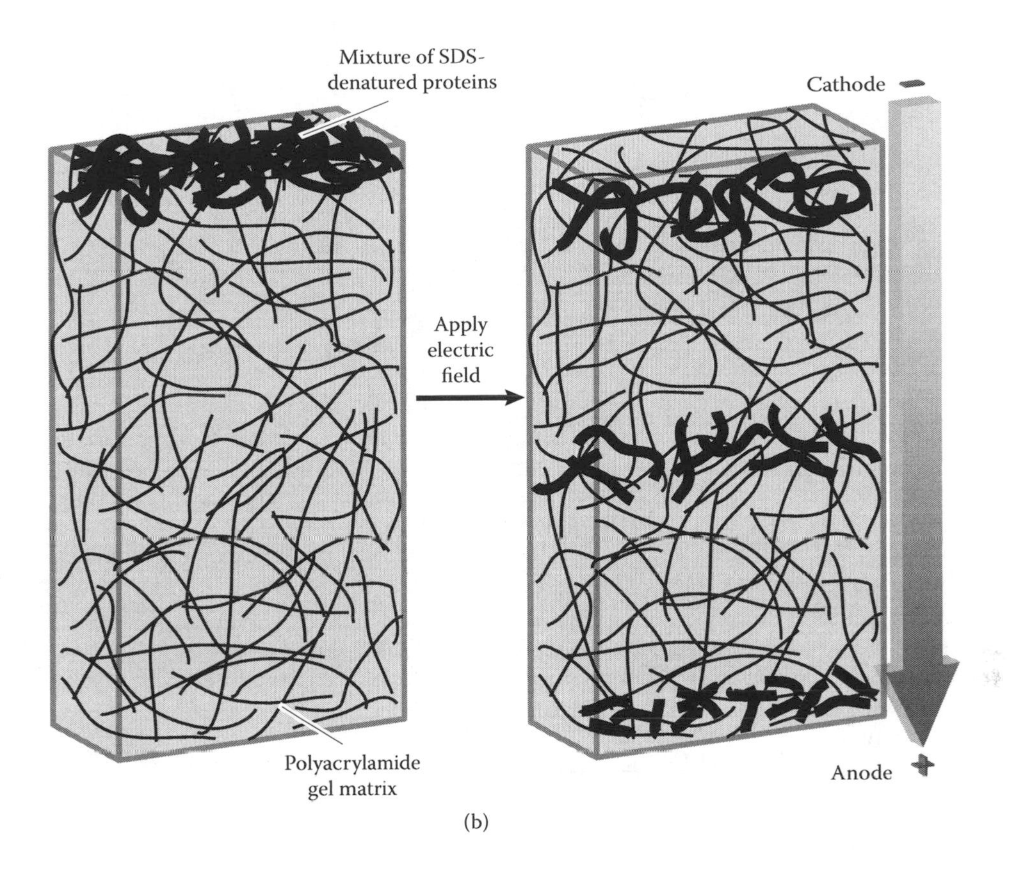

FIGURE 3.7 (continued)
(b) During electrophoresis, the SDS-treated polypeptide chains are sieved through a polyacrylamide matrix, and subsequently resolved by their molecular weights. The smaller proteins have higher mobility and thus move farther toward the anode.

Notes

1. The main steps for SDS-PAGE are outlined in Figure 3.8. We will use gels with discontinuous buffer systems.
2. Typical reagents used in SDS-PAGE and their functions are listed in Table 3.2.

Procedures

Part 1 (Day 1). Running the SDS-PAGE

1. *Check-in*
   a. Samples and materials
      - Checkpoint samples CP#1, CP#2, and CP#3 from Session 2
   b. Reagents
      - SDS-PAGE running buffer, 10X
      - Protein molecular weight standards
      - Sample buffer, 5X
      - 2.0 *M* NaOH

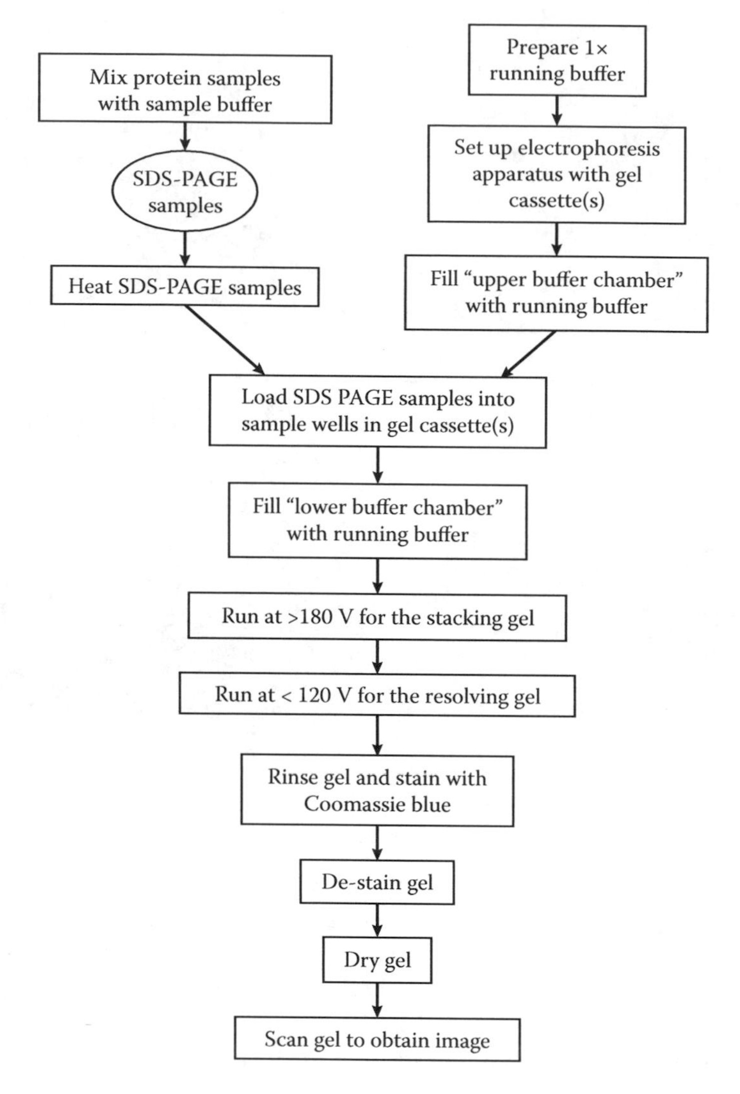

FIGURE 3.8
Outline of the main steps for running SDS-PAGE.

TABLE 3.2
Typical Reagents Used in SDS-PAGE and Their Functions

Reagent	Function
Sample Buffer (Laemmli Buffer)	
SDS	Denatures proteins; adds uniform negative changes on polypeptide chains
Glycerol	Adds viscosity and density to the solution and limits diffusion of the protein samples
Bromophenol blue	Adds a blue color to the solution; moves ahead of the proteins to indicate the front of electrophoresis
Tris-HCl[a]	Buffers the solution at pH 6.8; provides chloride ions as leading ions in the stacking gel (see "SDS-PAGE with Discontinuous Buffer Systems")
2-Mercaptoethanol	Breaks down disulfide bonds in proteins

continued

TABLE 3.2 (continued)

Typical Reagents Used in SDS-PAGE and Their Functions

Reagent	Function
Running Buffer	
SDS	Maintains the presence of SDS throughout the electrophoresis system
Tris-HCl[a]	Buffers the solution at pH 8.8
Glycine	Buffers the solution at pH 8.8; acts as trailing ions in the stacking gel
Staining Solution	
Coomassie blue	Stains proteins with dark blue color
Polyacrylamide Gel	
Polyacrylamide	Forms the gel matrix; its density partially determines the pore size of the gel
Bis-acrylamide	Cross-links polyacrylamide and partially determines the pore size of the gel
SDS	Maintains the presence of SDS
Tris-HCl[a]	Buffers the stacking gel at pH 6.8 and the resolving gel at pH 8.8
Other	
Molecular Weight markers	Mark the physical locations of proteins with different molecular weights on gel

[a] Other buffering reagents such as HEPES may be used in place of tris-HCl.

- Staining solution
- (Optional) Destaining solution

c. Special equipment and supplies
   - Gel-loading pipette tips
   - Precast 5% or 8% polyacrylamide gels
   - SDS-PAGE apparatus and power supply
   - Heating block
   - Shaker
   - Containers for gel staining and destaining

2. *[By instructor] Prelab preparation*: Turn on a heating block or a hot water bath and set the temperature close to 100°C. ***Caution:*** Hot surfaces!
3. *Preparing 1X running buffer*: Measure 900 ml of de-ionized water, and 100 ml of 10X running buffer. First pour the 10X running buffer into a 1000-ml bottle, and then rinse the cylinder twice with part of the 900 ml de-ionized water. Add the rinse and the rest of the de-ionized water into the 1000-ml bottle. Label the bottle as "1X running buffer."
4. *Sample preparation*
   a. Label three 0.5-ml or 1.5-ml microcentrifuge tubes as 1, 2, and 3. Prepare samples by mixing the following ingredients. *Note:* Vortex the tubes to ensure complete mixing.
      - Tube 1: 20 µl 5X sample buffer, 75 µl of CP#1; titrate in 2.0 *M* NaOH until the color turns dark blue (the same color as the 5X sample buffer). *Note*: The NaOH titration neutralizes the acetic acid in the sample.
      - Tube 2: 20 µl 5X sample buffer, 75 µl of CP#2; titrate in 2.0 *M* NaOH until the color turns dark blue.

- Tube 3: 20 µl 5X sample buffer, 20 µl of CP#3, 60 µl of 1X running buffer (from the "1X running buffer" bottle); titrate in 2.0 *M* NaOH until the color turns dark blue.

b. Heat the tubes with the heating block or by boiling for 10 minutes. *Note:* This treatment speeds up the denaturation of the tropocollagen molecules.

5. *Setting up the gel apparatus*

a. *Note:* The configuration of all PAGE apparatus is based on the same principle: an even electric field is applied across the gel through two platinum wires that are connected to the cathode and the anode, respectively (Figure 3.9). Observe your PAGE apparatus and follow the electricity from the cathode plug (usually colored black) to a platinum wire, through the gel (to be put in place later), to another platinum wire, and finally to the anode (usually colored red). Also identify the "upper buffer chamber," which is connected to the cathode, and the "lower buffer chamber," which is connected to the anode.

b. Open the package of a precast 5% or 8% SDS polyacrylamide gel and rinse the gel cassette with water. *Note:* Once the gel is opened, do not leave it in air for long

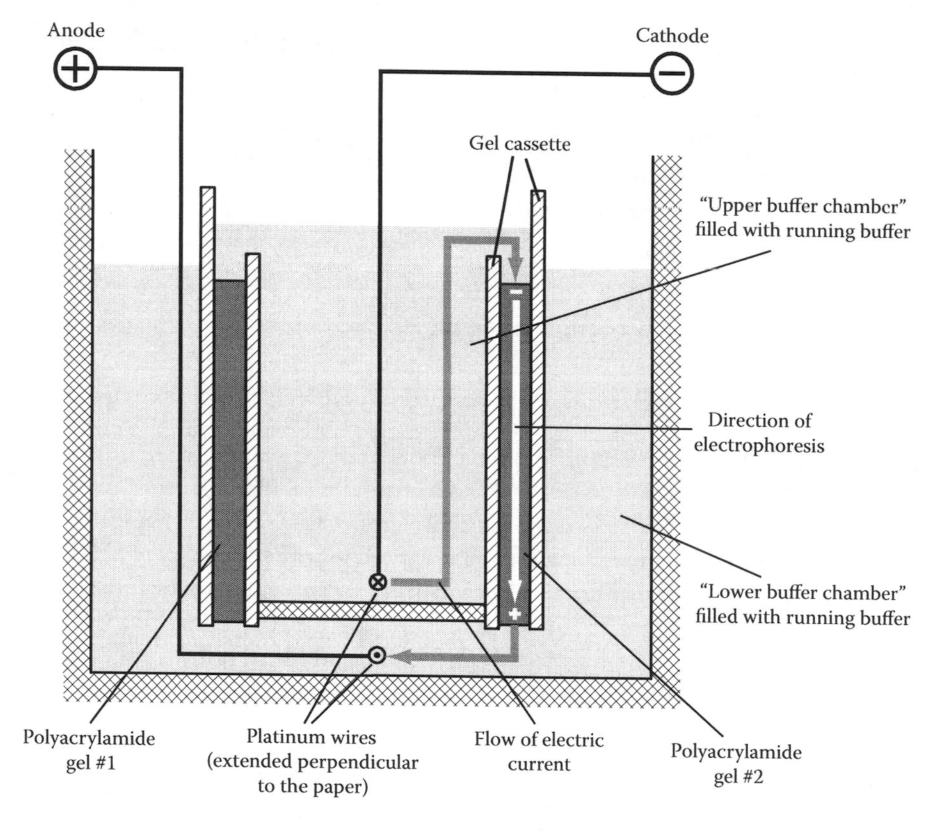

FIGURE 3.9
Cross-section view of a typical configuration for "submerged" gel electrophoresis. Two buffer chambers, the lower buffer chamber and the upper buffer chamber, are connected to the anode and the cathode, respectively, and the two chambers are connected only through the polyacrylamide gels. This configuration is termed "submerged" because the gel cassettes are in direct contact with the running buffer, which helps to remove the heat generated by the electric current. Two gel cassettes can be installed and run simultaneously; but if only one gel is run, a block needs to be placed in the other gel slot.

(>10 minutes) or the gel will begin to dry. If applicable, remove the "comb" that molds the sample wells. (Some precast gels do not have combs.) You can use a Sharpie marker to draw outlines of the wells to help with sample loading later.

c. Assemble the gel apparatus according to the user's manual, and pay attention to the following:
   - The sample loading wells should be connected to the upper buffer chamber.
   - The bottom of the gel should be connected to the lower buffer chamber.
   - If the gel apparatus has two gel slots, and only one gel is being run, make sure that the other slot is blocked.
   - Do not use excessive force to secure the gel cassette, or the plastic or glass plates might crack.

d. Add 1X running buffer to the upper buffer chamber. Check for leaks.

6. *Loading the samples*
   a. Take the samples off the heating block or hot water bath and let them cool for a few minutes. Centrifuge the tubes for 30 seconds using a minicentrifuge, and then vortex briefly to mix.
   b. Designate sample wells for the molecular weight marker and the samples. Use Table 3.3 as an example. (Each sample is loaded into multiple wells so that every group member should have a chance to practice the art of gel loading.)
   c. Sit or stand in front of the gel apparatus with a comfortable posture so that you can see the sample wells at eye level. Use *gel-loading tips* to pipette samples. To deliver sample into a well, place the gel-loading tip near the bottom of the well but *do not* penetrate the gel. (You might need to use both hands to hold the micropipette steady.) Slowly deliver the sample from the gel-loading tip. Since the sample is heavier than water, it will sink to the bottom of the well. Move the pipette tip up as you deliver the sample. Ideally, there should be minimal mixing between the sample and the running buffer.

7. *Running the gel*
   a. Add the 1X running buffer to the lower buffer chamber. Plug the apparatus to the power supply unit. Set the voltage to 120 V on the power supply and start the run. *Note:* Bubbles should arise from the cathode and the anode when the gel electrophoresis is running. The bubbles are H_2 on the cathode and O_2 on the anode, respectively, as a result of electrolysis of H_2O. (Do you see more bubbles at the cathode than the anode? Why?)
   b. The blue bands in the sample wells should start to collapse into a thin blue line. When the thin blue line passes the stacking gel, reset the voltage to 180 V. The run will take about ~30–40 minutes. The bromophenol blue molecules will

TABLE 3.3

Example of Lane Assignment on a Gel for the SDS-PAGE Samples

Lanes	Sample
1	15 µl molecular weigh marker
2, 5, and 8	20 µl of sample in tube 1
3, 6, and 9	20 µl of sample in tube 2
4, 7, and 10	20 µl of sample in tube 3

move to the front of the moving molecules and form what is called a "dye front," visibly the same thin blue line as seen before. ***Caution:*** Make sure that the dye front does not run out of the bottom of the gel.

c. Finish the run when the dye front is close to the bottom of the gel (~0.1–0.5 cm). To finish, first turn off the power supply, and then *unplug* the power supply leads from the unit. Discard the running buffer, and take out the gel cassette.

8. *Staining the gel*
   a. Open the gel cassette according to manufacturer's instruction. Use a razor blade to cut away the strips of gel that form the sample wells.
   b. Add de-ionized water to a gel-staining container. Rinse the gel (still on a plate at this point) with de-ionized water, then lift one of corners at the bottom of the gel and gently peel the gel off. Place the gel in the container. Another way to lift the gel is to hold the plate flat over the container but with the gel facing down, and then carefully peel down one corner of the gel until it falls into the container by its own weight. Rinse the gel by gently shaking it for 3 minutes, and change the de-ionized water to repeat the rinse twice more.
   c. Discard the last rinse and drain (or pipette) off the water, then add enough staining solution to submerge the gel. Place the container on the shaker and let the gel stain overnight.
9. *[By instructor] Destaining the gel:* Discard the staining solution into a proper waste container. Rinse the gel with de-ionized water, and then add destaining solution or just use de-ionized water to destain the gel overnight.
10. *Finishing up:* Rinse the gel apparatus with water. Properly discard the micro-centrifuge tubes and the gel cassette.

SDS-PAGE with Discontinuous Buffer Systems

SDS-PAGE with discontinuous buffer systems is the most commonly used method for protein analysis. A discontinuous gel has two portions: the top portion, usually with sample wells molded in, is called the stacking gel, and the bottom portion is called the resolving gel. The differences between the stacking gel and the resolving gel are:

- The stacking gel typically contains 4% polyacrylamide with a low degree of cross-linking, which makes it highly porous. The polyacrylamide content of the resolving gel ranges from 5% to 20%, or contains a gradient of percentages, depending on the molecular weights that need to be resolved.
- The pH of the stacking gel is typically 6.8, and that of the resolving gel is typically 8.8.

Discontinuous systems are used to achieve high resolution in protein separation. When the protein samples are first loaded into the sample wells, the heights of the sample bands depends on the sample volumes (Figure 3.10a). These bands need to be "squeezed" into thin lines for fine resolution of the proteins, and that is exactly the function of the stacking gel. When the electric field is turned on, the Cl^- ions, being negatively charged and small, will move quickly to the front of the proteins in the stacking gel. These Cl^- ions are therefore the *leading ions*. Following behind are the denatured and negatively charged proteins, and trailing behind the proteins are the glycine molecules, which serve as *trailing ions*. The low electromobililty of the glycine molecules is due to the fact that they are barely negatively charged at pH 6.8, the pH of the stacking gel. As the electrophoresis progresses through the stacking gel, the strong voltage drop between the leading ions and the trailing ions (i.e., glycine) drives the protein sample bands into thin lines (Figure 3.10b). The large pores in the stacking gel allow free passage to all protein molecules, which helps to "line up" these molecules before they enter the resolving gel. In the resolving gel, which has a pH of 8.8, the glycine molecules become completely negatively charged, and they move at a similar speed as the Cl^- ions. As a result, the voltage drop that drives the proteins through the stacking gel disappears, and the protein molecules are separated into thin bands according to their molecular weights (Figure 3.10c).

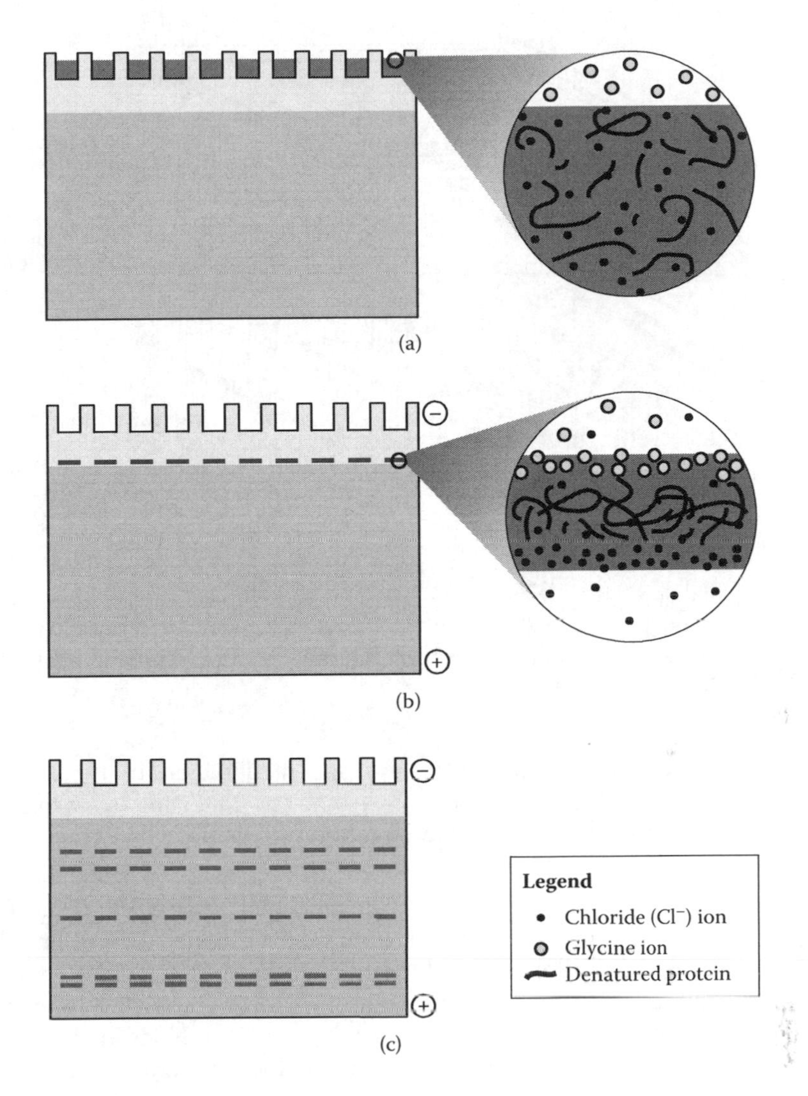

FIGURE 3.10
SDS-PAGE with discontinuous buffer systems.

Part 2 (Day 2). Drying the SDS-PAGE Gel

1. *Check-in*
   a. Samples and materials
      - Destained gel from Part 1
   b. Reagents
      - Gel-drying solution
   c. Special equipment and supplies
      - Cellophane sheets
      - A set of gel-drying frame: 1 frame, 1 solid back plate, and 4 side clips
2. *Preparing the gel:* Replace the destaining solution (or water) with gel-drying solution. Shake the gel for 10 minutes. *Note:* This step allows exchange of the liquid in

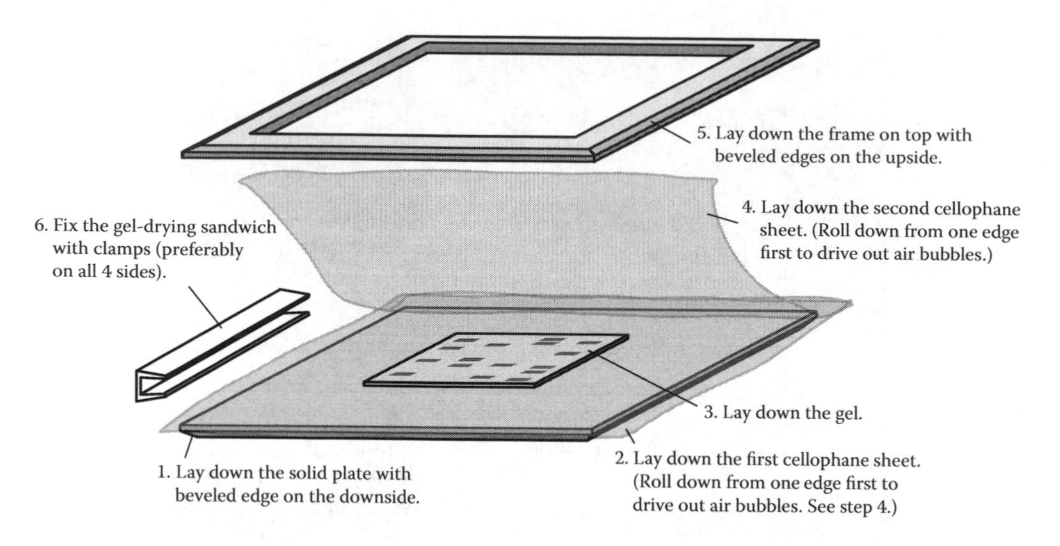

FIGURE 3.11
Drying a polyacrylamide gel.

the gel to the gel-drying solution, which helps to reduce swelling and makes the gel more flexible.

3. *Framing the gel*
   a. Procedures for framing a gel for drying are depicted in Figure 3.11. First hydrate two cellophane sheets in a pan with tap water for a few minutes. Lay down a few paper towels, and then lay down the solid plate with the beveled edges on the down side.
   b. Lay down one cellophane sheet on the solid plate by rolling it out from one edge, and smooth out any air bubbles trapped in between. Add some gel-drying solution on top of cellophane, and then carefully place the gel on it. If air bubbles are trapped, add more gel-drying solution to the surface and try again. Make sure that the gel is spread out and that no air bubbles are trapped. (*Note:* If the gel is broken, just assemble the pieces in the right place on the cellophane sheet.) Add more gel-drying solution on the gel, and lay down the other cellophane sheet. Again, make sure that no air bubbles are trapped.
   c. Now place the open frame on top with its beveled edges on the up side. Slide in the four side clips to bind the frame and the plate. Label the set with a piece of tape.
   d. Allow the gel to dry in air, or preferably, in a fume hood. The drying time varies from overnight to a couple of days depending on the ambient conditions.
4. *Finishing up:* Pour the gel-drying solution to a waste container. Rinse the gel container and place it on paper towels to dry. Paper towels soaked with gel-drying solution can be discarded into regular trash.

Part 3 (Day 3). Imaging the Dried Gel

Remove the dried gel from the gel-drying frame. Trim away the cellophane around the gel. You can place the gel inside your notebook to keep it flat. Scan the gel on a scanner and save the gel image for your lab report.

Session 4. Fabricating Collagen/Chitosan Sponges

Soluble and insoluble collagen or their combinations are optimized for different biomedical applications. For example, as injectable tissue filler, soluble collagen is used for smoothing small wrinkles, but a mixture of soluble and cross-linked collagen is used for deep furrows because it can provide stronger support. For another example, highly cross-linked collagen is used for wound dressings or hemostatic dressings because these applications require stronger mechanical properties and a slower degradation rate. Therefore, it is important to study the chemical and physical properties of insoluble, insoluble collagen. In this session, we will focus on the processing of cross-linked collagen by fabricating collagen sponges using procedures outlined in Figure 3.12. For convenience, we will again use calf skin as a starting material (although mature bovine skin is used for commercial production because of its higher degree of collagen cross-linking, better availability, and lower material cost). The calf skin is first treated with a highly basic and reducing solution to facilitate the removal of the epidermal portion of the skin, this depilation process also helps to destroy noncollagen proteins and reduce the antigenicity of the final product. The dermis portion of the skin (also know as the corium) is then acid-swollen so that it can be homogenized easily. The homogenzied pulp contains a mixture of soluble and insoluble collagen. Subsequently, the pulp is then frozen in molds, and after freeze-drying, porous, absorbent sponges will form. Two examples of collagen and collagen/chitosan sponges are shown in Figure 3.13. While the porosity of the sponges depends on the collagen-to-water ratio in the pulp, the pore size depends largely on the freezing rate: faster freezing results in smaller ice crystals in the frozen pulp, and subsequently smaller voids after the ice crystals are removed through freeze-drying. The sponges can be further strengthened by dry heat treatment, which increases the degree of cross-linking in collagen; such treatment can serve as a sterilization process as well.

Chitosan, another natural biomaterial, can also be fabricated into sponge through freeze-drying. Since chitosan is soluble in acid, it can be readily mixed with collagen in a suspension; thus, collagen/chitosan composite sponge can be fabricated in the same way as pure collagen sponge.

Procedures

Part 1 (Day 1). Preparing the Calf-Skin Dermis (Corium)

1. *Check-in*
   a. Samples and materials
      - Fetal calf skin (or adult bovine skin)
      - Chitosan powder
   b. Reagents
      - 0.5 *M* acetic acid
      - Na_2S solid
      - NaOH solid
   c. Special equipment and supplies
      - Dissection toolkit
      - 50-ml conical tube

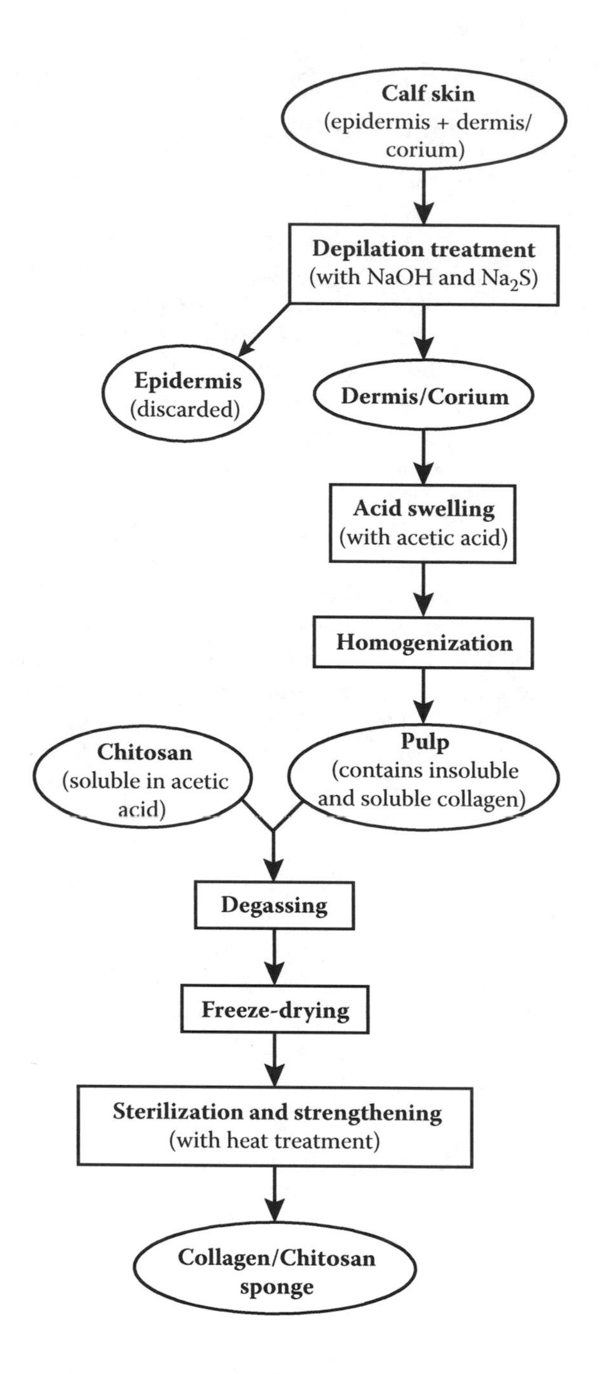

FIGURE 3.12
A flow chart of the procedures for fabricating collagen/chitosan sponge. Calf skin and its derivatives (in ovals) are processed with multiple steps (in rectangles) to produce collagen/chitosan sponge.

2. *Removing the epidermis*

    a. *Note:* The following treatment is based on Huc and Gimeno (1994). It might be best for the instructor to perform this step ahead of time or as a demonstration.

    b. Trim off fat tissues from the calf skin and wash clean with regular tap water. Cut the calf skin into ~3″ × 3″ pieces.

(a)

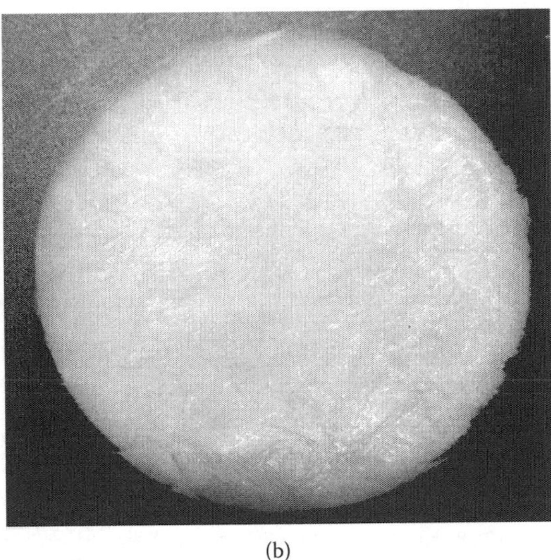

(b)

FIGURE 3.13
Examples of collagen sponges. (a) A commercial hemostatic sponge. (b) A collagen/chitosan sponge made in laboratory freeze-drying.

c. In a 1000-ml beaker, add 500 ml de-ionized water and dissolve ~5.2 g sodium sulfide and ~7.0 g sodium hydroxide in it. Soak five to seven cut pieces of calf skin in this solution for at least 48 hours. *Note:* This step helps to break down the epidermis and facilitates its removal.

d. Use a scalpel or a razor blade to shave or peel off the epidermis; the dermis, or the corium, should have a translucent white color (Figure 3.14).

e. Wash the depilated calf-skin extensively with tap water, and then soak it twice in de-ionized water for at least 24 hours each to remove residual Na_2S and NaOH.

f. After the depilation treatment, the corium pieces can be used right away or frozen for later use.

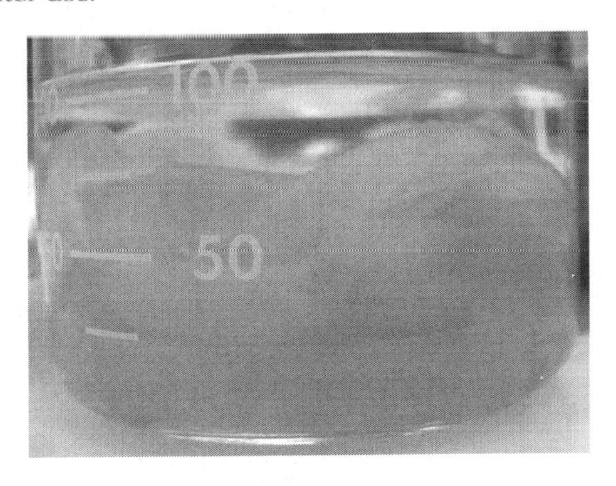

FIGURE 3.14
Calf-skin dermis (corium) after removal of the epidermis.

3. *Acid swelling of calf-skin dermis/corium*
   a. Dry a piece of calf-skin dermis on tissue paper. Use a scalpel or a pair of scissors to cut it into small pieces of ~0.5 cm × 0.5 cm.
   b. Weigh ~4 g of corium pieces and place them in a 50-ml conical tube. To make collagen/chitosan composite sponge, weigh 0.1 g of chitosan and add to the tube. (The chitosan in the composite will be between ~10% and 20% by weight). Add 40 ml of 0.5 *M* acetic acid to swell the collagen. The chitosan will be dissolved as well. Label the tube, and store it at 4°C for 48 hours or longer.

Part 2 (Day 2). Making Collagen/Chitosan Sponges

1. *Check-in*
   a. Samples and materials
      – Acid-swollen calf-skin dermis/corium, optionally with chitosan
   b. Special equipment and supplies
      – Ice and ice bucket
      – Homogenizer
      – Degassing chamber
      – Mold (6-well polystyrene plate)
      – Freeze-dryer
2. *Observation:* The calf-skin corium should be swollen, the chitosan solid should have dissolved, and the mixture should be thick and viscous.
3. *Homogenizing the acid-swollen corium*
   a. Set the 50-ml conical tube that contains the acid-swollen corium firmly inside a bucket (or beaker) of ice. Insert the generator probe of the homogenizer into the middle of the tube; make sure that the hole in the probe is submerged. Allow the setup to cool for a few minutes in ice.
   b. Repeat the following procedures, and allow 10 seconds in between each homogenization for the sample to cool: Turn the homogenizer to the highest power for 5 seconds, shut if off, and move the tube up and down a few times to dislodge any big chunks of corium. Repeat until a viscous pulp has formed and no large chunks of corium are visible.
   c. (Set up Step 4 first.) To clean the homogenizer probe, fill a large beaker with tap water and run the probe in it with several changes of the water. After the probe is cleaned, rinse it with de-ionized water. Unplug the homogenizer and wipe the probe dry with tissue paper.
4. *Degassing:* Transfer the collagen/chitosan pulp to a 100-ml beaker and place it in a bell jar vacuum that is connected to a water aspirator pump. Turn on the water and allow the sample to degas for 15 minutes or until there is little air bubble in the pulp. (You can now go back to Step 3c.)
5. *Freezing:* Transfer the degassed collagen/chitosan pulp into the wells of a 6-well plate. Replace the lid cover and place the plate in a –20°C freezer. This allows the pulp to freeze at a relatively rapid speed, which results in smaller pore sizes in the sponge. Let the pulp freeze overnight.

6. *Freeze-drying:* Your instructor will put the frozen collagen or collagen/chitosan pulp into the freeze-dryer for you the next day. After the freeze-drying is finished, remove the collagen/chitosan sponges from the mold with a pair of tweezers and observe the sponge for its color, porosity, and strength.

(Optional) Part 3 (Day 3). Heat Treatment of the Collagen Sponge

Incubate one collagen/chitosan sponge in a dry oven set at 110°C for 48 hours. Observe the sponge again after the heat treatment and compare it with the untreated sponge.

Questions

1. Type I tropocollagen has a length of 3000 Å and a diameter of 15 Å. Which one of the following is shaped like an extended tropocollagen collagen the most?
   a. A no. 2 pencil that is 7 1/4 in. long and 3/8 in. in diameter.
   b. A 1.5 cm clip of a medium-sized human hair that is about 80 μm in diameter.
   c. A flagpole that is 80 ft long with an average diameter of 8 in.
2. What is the reason to cryo-pulverize calf skin before extraction (besides a lot of fun)?
3. Why do we use fetal calf skin to extract acid-soluble collagen? Can we use adult bovine skin?
4. In Session 1, the collagen that we extracted is at which level of the collagen structural hierarchy?
   a. Single polypeptide strand
   b. Tropocollagen
   c. Collagen fibril
   d. Collagen fiber
5. In Session 1, when setting up acid extraction of ~5 g of pulverized fetal calf skin, you were instructed to use 160 ml of 0.5 *M* acetic acid. What should you do if you discovered after the extraction is completed that you had only added 150 ml of 0.5 *M* acetic acid?
6. In Session 2, to salt-precipitate type I collagen, you added NaCl to a final concentration of 1.0 *M*. If you accidentally added more NaCl and the final concentration became 1.5 *M*, how would the purification process be affected? If you discovered the mistake too late, and you had already centrifuged the precipitated collagen and discarded the supernatant, how would you remedy the situation?
7. There are hundreds of different proteins in blood plasma. A coagulation protein, fibrinogen, is circulated in the blood at a relatively high concentration of ~2.6 mg/ml. To purify this protein from the plasma, which of the following methods is the best *first* step? Why?
   a. Dialysis
   b. Gel electrophoresis
   c. Precipitation by ammonia sulfate or sodium chloride
   d. Extraction with 0.5 *M* acetic acid

8. A protein you are working with is in a solution of 100 *mM* of NaCl and 10 *mM* phosphate, but for a particular experiment, you need to change the solution to 100 *mM* KCl and 50 *mM* phosphate. How would you do it?
9. In SDS PAGE, the proteins are resolved mainly by:
   a. The numbers of negatively charges amino acids, i.e., glutamate and aspartate, in the proteins
   b. The amino acid sequences of the proteins
   c. The numbers of amino acids in the proteins
10. When fabricating collagen sponge using freeze-drying, how does the rate of temperature decrease during freezing affect the pore size of the sponge?
11. What are the similarities and differences between the methods for producing soluble collagen and insoluble collagen?

Appendix. Recipes and Sources for Equipment, Reagents, and Supplies*

Session 1. Extracting Acid-Soluble Type I Collagen from Bovine Calf Skin

- **Calf skin.** Pel-freez Biologicals, cat. no. 57090-1. *Sel. crit.:* Young animal skin.
- **Cryogenic pulverizer.** BioSpec Products, BioPulverizer, cat. no. 59014H. *Sel. crit.:* Cryogenic; capable of handling >1 g of sample each batch.

Session 2. Purification of Extracted Collagen

- **Dialysis tubing.** Regenerated cellulose, MWCO 6000 to 8000, vol./cm = 3.27, Fisher cat. no. 21-152-4. *Sel. crit.:* Made with regenerated cellulose, MWCO <10,000, vol./cm ratio ~3.0–3.5.
- **Centrifuge, rotor, and tubes.** See Module I appendix (Chapter 2).
- **Freeze-dryer.** See Module I appendix (Chapter 2).

Session 3. Analysis of Collagen Extraction and Purification by Electrophoresis

- **SDS-PAGE running buffer.** 10X Tris/Glycine/SDS electrophoresis buffer, Bio-Rad Laboratories, cat. no. 161-0732EDU. *Sel. crit.:* 30.3 g Tris base, 144 g glycine, and 10 g SDS in 1 L H_2O.
- **Protein molecular weight standards.** Precision Plus Protein Kaleidoscope Standards, Bio-Rad Laboratories, cat. no. 161-0375EDU. *Sel. crit.:* Pre-stained.
- **Sample buffer, 5X (10 ml).** 1.2 g SDS, 5.0 ml glycerol, 2.5 ml of 1.0 *M* tris-HCl pH 6.8 stock solution, 0.5 mg Bromophenol blue, 0.5 ml β-mercaptoethanol, add H_2O to 10 ml.

* *Disclaimer*: Commercial sources for reagents listed are used as examples only. The listing does not represent endorsement by the author. Similar or comparable reagents can be purchased from other commercial sources. See selection criteria (sel. crit.).

- **Staining solution.** Bio-Safe Coomassie stain, Bio-Rad Laboratories, cat. no. 161-0786EDU. *Sel. crit.:* Coomassie blue based.
- **De-staining solution.** *Sel. crit.:* Follow instruction on user's manual for the staining solution.
- **Gel-loading pipette tips.** Fisherbrand gel-loading tips, Fisher cat. no. 02-707-138. *Sel. crit.:* Capillary tipped.
- **Precast polyacrylamide gels.** Tris-HCl Ready Gel Precast Gel, 10-well, 30 µl, Bio-Rad Laboratories, cat. no. 161-1104EDU (4–15% linear gradient) or 161-1210EDU (5% resolving gel). *Sel. crit.:* Acrylamide percentage >5% but >10%; gels must fit the apparatus.
- **SDS-PAGE apparatus and power supply.** Protein Electrophoresis Classroom Lab Set, Bio-Rad Laboratories, cat. no. 166-0810EDU, 166-0820EDU, or 166-0801EDU. *Sel. crit.:* Submerged vertical electrophoresis system with buffer tank.
- **Heating block.** Fisher Scientific Isotemp dry bath incubators, 1-Block, Fisher cat. no. 11-718Q. Block cat. no. 11-715-307Q. *Sel. crit.:* Heats up to 100°C; block must fit 1.5-ml or 0.5-ml microcentrifuge tubes.
- **Gel-drying solution (100 ml).** 4 ml glycerol, 76 ml water, and 20 ml ethanol (4% v/v glycerol, 20% v/v ethanol).
- **Gel-drying frame and cellophane sheets.** Thermo Scientific Owl gel-drying kits 10 cm × 10 cm, mfr. no. GDF-10, Fisher cat. no. OWGDF10. *Sel. crit.:* Match the size of the gel-drying frame with the cellophane sheets.

Session 4. Fabricating Collagen/Chitosan Sponges

- **Chitosan.** Sigma-Aldrich cat. no. 448877. *Sel. crit.:* Soluble in acid.
- **Dissection toolkit.** Fisherbrand dissecting set, Fisher cat. no. 08-855. *Sel. crit.:* Similar components.
- **Degassing chamber.** Nalgene vacuum chamber, mfr. no. 5305-0609EMD, cat. no. S413811. *Sel. crit.:* Capacity >4 L.
- **Water aspirator pump.** Airejector aspirator, Fisher cat. no. S41380. *Sel. crit.:* Threaded to fit water tap in your lab.

References

Bella, J., M. Eaton, B. Brodsky, and H.M. Berman, Crystal and molecular structure of a collagen-like peptide at 1.9 Å resolution, *Science*, 266, 75–81, 1994.

Dutta, P.K., M.N.V. Ravikumar, and J. Dutta, Chitin and chitosan for versatile applications, *J. Macromol. Sci. C-Pol. Rev.*, 42, 307–354, 2002.

Heino, J., The collagen family members as cell adhesion proteins, *BioEssays*, 29, 1001–1010, 2007.

Huc, A., and R. Gimeno. *Process of Preparation of Collagen Containing in Major Proportion Insoluble Collagen and Collagen Having High Mechanical Resistance and Thermal Stability Obtained Thereby*, Coletica, Lyons, France, U.S. Patent Office, Washington, D.C., 1994.

Lakes, R., Materials with structural hierarchy, *Nature*, 361, 511–515, 1993.

Lee, C.H., A. Singla, and Y. Lee, Biomedical applications of collagen, *Int. J. Pharm.*, 221, 1–22, 2001.
Sun, Y.-L., Z.-P. Luo, A. Fertala, and K.-N. An, Direct quantification of the flexibility of type I collagen monomer, *Biochem. Biophys. Res. Commun.*, 295, 382–86, 2002.

4

Basic Laboratory Skills II. Cell Culture

I Cell Culture

Cell culture is the process of growing cells in a controlled, *in vitro* environment in the laboratory. The cells can be either isolated directly from tissues or continued from previously cultured cell lines. Cells in cultures can be systematically studied, manipulated, and engineered. The development of cell culture (or tissue culture) has played a pivotal role in the progression of molecular biology and cell biology, as the vast amount of knowledge that we have today about cellular activities and functions comes from studies of cells in cultures. Cell culture has also become a fundamental part of biomaterials and bioengineering; for example, in biomaterials studies, we use cultured cells to study cell-material interactions, and in tissue engineering, we use cultured cells to construct artificial organs.

Cell culture is such a versatile method partly because many types of cells, from skin cells to stem cells, can be isolated from various tissues and at various development stages, and successfully cultured *in vitro*. When cells are freshly isolated from tissues and have not proliferated extensively *in vitro*, the culture is a *primary culture*. At this stage the cells mostly retain their *in vivo* characteristics and activities. When they are allowed to divide and proliferate, the cells may undergo morphological and genetic changes and become what is called a *secondary culture*. After a certain number of generations, most types of cells will stop dividing and enter a senescence state. However, some cells can either spontaneously or be induced to transform into "immortal" cells—that is, cells that can keep on dividing; the cells have then become a *cell line*. Cells of a certain cell line can be very different from their *in vivo* counterparts, but they usually retain some activities that are similar to the latter's. For example, an important function of fibroblast cells in the dermis tissues of skin is to secrete collagen, a major component of the extracellular matrix that provides support for the tissues and contributes largely to the mechanical properties of skin; two fibroblast cell lines, L-929 and NIH 3T3, retain this function and can be induced to secrete collagen in cultures. (In our experiments, L-929 cells will be used for biocompatibility testing, and NIH 3T3 cells for growing artificial skin.)

Cell cultures must be maintained in a carefully controlled environment that mimics the physiological environment in terms of temperature, nutrients, pH, sterility, and so forth. The following are some of the most critical features for a cell culture environment:

- **Temperature.** The most commonly used temperature for mammalian cell cultures is 37°C, a temperature that is close to the body temperatures of most mammalian organisms.
- **Medium.** Cell growth media contain the salts, amino acids, vitamins, carbohydrates, growth factors, etc. that are essential nutrients for the survival and growth of cells.

- **pH.** A stable pH of ~7.4 is commonly used for cell cultures, and this pH is maintained by the buffering reagents in growth media as well as the level of CO_2 in the atmosphere inside the controlled environment.
- **Maintenance.** To maintain the level of nutrients and remove wastes, culture media need to be renewed on a regular basis.
- **Sterility.** Cells need to be cultured in an environment that is free of microbial infection since they no longer have the benefit of a defense system that is part of the physiological environment. Therefore, reagents, culture vessels, etc. need to be in strict sterile conditions, and aseptic techniques must be used when working with cell cultures.

In the following lab exercises, we will first familiarize ourselves with basic tissue culture equipment and supplies, and then we will learn some of the basic cell culture techniques, including aseptic techniques.

I.1 Equipment for Cell Culture

The following is a list of typical equipment used in a tissue culture facility:

- **Biosafety cabinet (BSC).** A biosafety (or biological safety) cabinet is a contained workspace that provides a sterile environment for handling biological samples as well as protection for the worker from the samples' potential hazards. All BSCs use high-efficiency particulate air (HEPA) filters to remove airborne particulates and aerosols larger than 0.3 μm, thus eliminating most airborne microbes, and the filtered air is effectively sterile. Tissue culture laboratories that work with biosafety levels 1, 2, and 3 samples are typically equipped with class II BSCs. Figure 4.1 shows a schematic of a typical airflow pattern in class II BSCs. In addition, BSCs are usually equipped with germicidal UV lamps to eliminate potential microbial contamination when they are not in use. (The UV lamp must be turned off when the BSC is in use.)

Biosafety Levels

Biosafety refers to the safe handling and transfer of biological agents, which can be bacteria, recombinant DNA, human blood, etc.; for cell culture labs, biosafety mainly concerns the safe handling of cells. Note that the safety emphasis is on the lab workers who work with cells (as opposed to protecting cells from being infected, which is what aseptic techniques are for). Can cells harm us? The answer is: yes, potentially. Cells can harbor the same viruses—such as hepatitis, HIV, and other microbes—that can infect us; thus, it is important to prevent transmission of any potential infectious agents from the cells to the lab workers. Safety measurements include applying special practices in the lab, using safety equipment such as BSC, and working in a properly equipped facility. Several U.S. government agencies have published biosafety guidelines that categorize biosafety into four levels. The Centers for Disease Control (CDC) have published a document entitled "Biosafety in Microbiological and Biomedical Laboratories" (BMBL) (4th edition) with definitions of the four biosafety levels and recommendations on lab practices, equipment requirements, and facility requirements. Your cell culture lab should be properly established following the BMBL guidelines and additional institutional guidelines. In our experiments, the three mouse cell lines L-929, NIH 3T3, and MC3T3-E1 are all biosafety level 1 cell lines, but the human cell line HaCaT is a biosafety level 2 cell line. To ponder: What is the biosafety level of calf skin, the kind that we used for collagen purification?

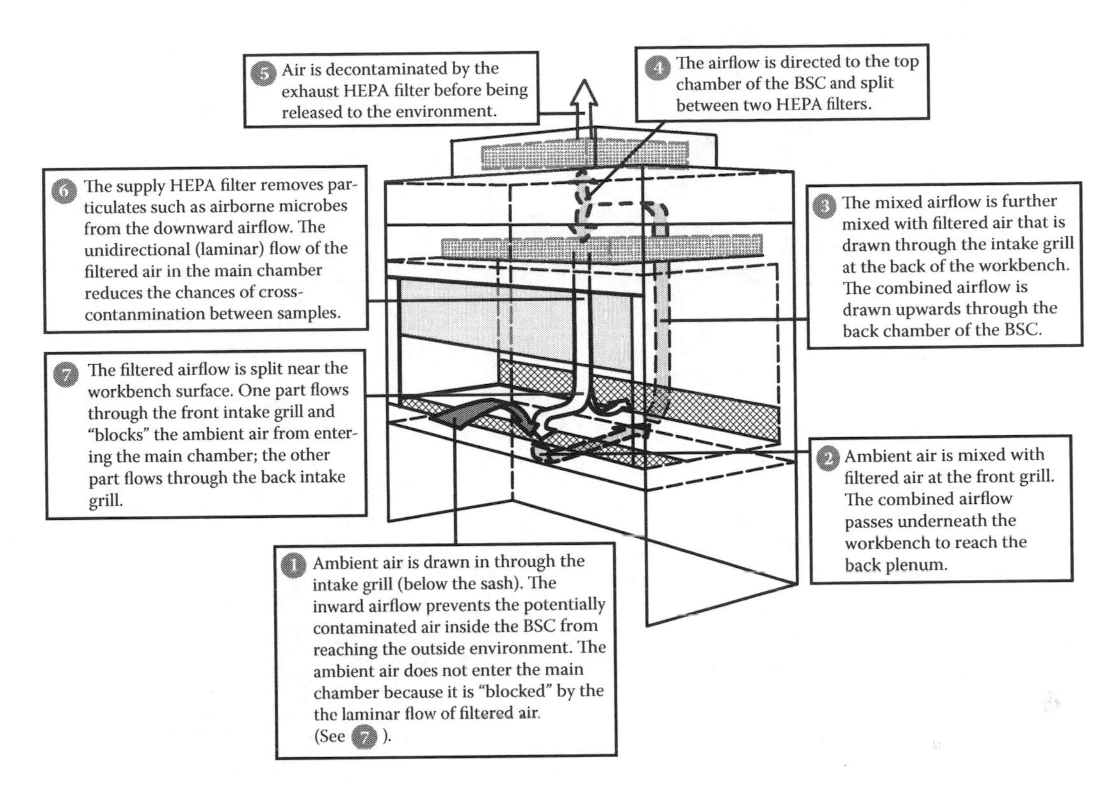

FIGURE 4.1
Typical air-flow scheme in a class II BSC.

- **Aspiration.** This is a BSC accessory that is used for efficient liquid removal. Typically a vacuum pump or house vacuum is used for suctioning liquid. The vacuum is sequentially connected to a liquid trap and a reservoir; the reservoir, typically a glass vacuum flask, is in turn connected to one of the built-in inlets of the BSC. Sterile, disposable glass (Pasteur) pipettes can be attached to this vacuum inlet through a hose. When vacuum is applied, liquid can be aspirated into the reservoir and decontaminated there by bleaching (Figure 4.2).
- **Pipet-aids.** See Basic Laboratory Skills I (Chapter 1). Specifically for cell culture, sterile serological pipettes are used, and the mouthpiece of the pipet-aid and the filter inside can be sterilized as well.

Why Is My Pipet-Aid Running but Not Taking Up Liquid?

This might happen to you one day if it has not already: You insert a serological pipette in a pipet-aid and press the "up" button to pipette, but the liquid is not being taken up even though you can hear that the motor inside the pipet-aid is running just fine. You push the pipette firmly into the mouthpiece of the pipet-aid for a good seal and try again, but still, no liquid is taken up at all. What's wrong? The most likely reason is a clogged filter in the mouthpiece—someone (or you!) was not paying attention and probably aspirated cell growth media past the pipette and soaked the filter inside the mouthpiece. When the protein-rich media dried, a film had formed over the pores of the membrane in the filter, effectively blocking it. As a result, even though the motor of the pipet-aid is working, air cannot be drawn across the filter, and so no liquid can be taken up through the pipette. To fix this problem, simply twist open the mouthpiece of the pipet-aid and replace the clogged filter. This situation can be avoided if you are careful and make sure that no liquid *ever* gets inside the mouthpiece of the pipet-aid. However, spare filters should always be kept around the lab just in case.

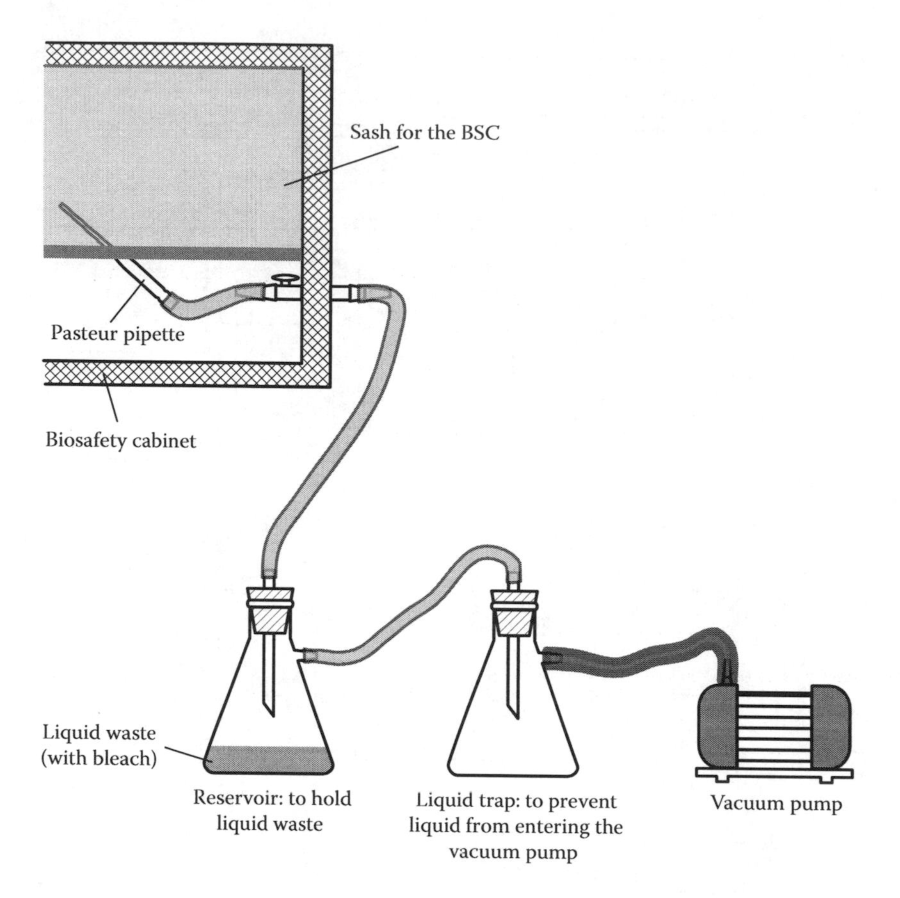

FIGURE 4.2
Schematic for setting up an aspiration device for a BSC.

- **Micropipettes.** See Basic Laboratory Skills I (Chapter 1). Specifically for cell culture, fully or partially autoclavable micropipettes are preferred.
- **CO_2 incubator.** A CO_2 incubator provides a controlled environment that resembles the *in vivo* physiological environment for cell cultures. Typical features for CO_2 incubators include precise temperature control, high relative humidity maintenance, CO_2 level control, and HEPA-filtered air circulation to remove particulates in the air. A typical CO_2 incubator is equipped with a thermostat and is water-jacketed to provide even heat distribution. The interior is usually made of stainless steel to prevent corrosion and to allow easy disinfection. A typical set of conditions that is used for many types of cells is: a CO_2 level of 5%, a humidity of 95%, and a temperature of 37°C.
- **Autoclave (pressurized steam sterilizer).** An autoclave is used to sterilize equipment, materials, and reagents with high temperatures (typically 120°C) and high pressures (typically 15 psi). Actual setting of temperature and duration for autoclaving may vary for different materials. It is important to note that (1) autoclaving of liquid requires slow exhaustion of the steam, and (2) materials must be heat-stable beyond 120°C to be autoclaved.

- **4°C refrigerator, –20°C Freezer, and –80°C Freezer.** Tissue culture reagents are stored refrigerated or frozen at appropriate temperatures. Typically the 4°C refrigerator is for short-term storage of thawed reagents or the ones that should not be frozen, whereas –20°C and –80°C freezers are for long-term storage. Follow instruction on storage condition for a given reagent. In particular, it is important to minimize the number of freeze-thaw cycles when handling labile reagents such as enzymes and sera.
- **Liquid nitrogen storage.** This is also called cryo-storage, and it is typically in the form of a vacuum-insulated tank partly filled with liquid nitrogen. A cryo-storage is typically used for long-term storage of frozen cells; the very low temperature of liquid nitrogen (–196°C) is crucial to maintaining the viability of the cells. Note that the frozen cells are typically stored "dry" in the vapor phase of the liquid nitrogen instead of fully submerged in it.
- **Centrifuge.** Cells in suspension can be "collected" into a pellet through centrifugation at low speed (<1000 ×g), and after removal of the supernatant, the cells can be resuspended in fresh medium or solution. This procedure can be used to change the concentration of the cells or to wash off a reagent, for example. In tissue culture labs, clinical centrifuges are commonly used for centrifugation of cells.
- **Water bath.** A water bath set to 37°C is necessary for thawing and warming up reagents and cells. The bath water should be changed from time to time to prevent overgrowth of microbes. *Important:* Bottles, tubes, etc. should not be submerged below the "neck" in water bath; otherwise, the reagents inside will risk being contaminated.
- **Inverted microscope and image capture.** An inverted microscope is an optical microscope that is configured specially for observing live cells that adhere to the bottoms of culture vessels. It is typically equipped with phase contrast, which allows observation of unstained, live cells. Images from the inverted microscope can be captured with a camera that is attached to the microscope. These images can be saved on a computer and analyzed later.

1.2 Consumable Items

Because of the strict requirement for sterility, and to prevent cross-contamination, many supplies used for cell cultures are disposable. These consumable items are sterilized by different means depending on the materials. Often these items are available with different sizes, accuracy, etc. to suit different experimental needs. Because of the presence of biohazards, proper disposal of these consumable items is strictly required. (See Section 1.3, Waste Disposal.)

- **Serological pipettes.** These pipettes are typically used with pipet-aids to handle sera, growth media, buffer, and other aqueous solutions with volumes ranging from ~ 0.1–100 ml. *Materials:* Commonly polystyrene. *Sterilization:* γ-radiation. *Disposal:* Biohazard waste or designated biohazard pipette container.
- **Pipette tips.** Pipette tips are used together with micropipettes to handle aqueous liquid as well as some organic solvents due to the better chemical resistance of the polypropylene in comparison to polystyrene. *Materials:* Commonly

polypropylene. *Sterilization:* Autoclave (γ-radiated packs are often available, too). *Disposal:* Biohazard waste.

- **Pasteur pipettes.** These are most commonly used for liquid aspiration. *Materials:* Glass. *Sterilization:* Autoclaved in stainless steel or polycarbonate canister. *Disposal:* Biohazard *sharps* container.
- **Tissue culture vessels.** Tissue culture vessels come in a variety of shapes and sizes, which include Petri dishes, flasks, multiwell plates, etc. The cell growth surface of these vessels are usually specially treated (by oxygen plasma, for example) to enhance cell attachment. Be sure to use only vessels that are designated for tissue culture use. *Materials:* Commonly polystyrene. *Sterilization:* γ-radiation. *Disposal:* Biohazard waste.
- **Conical centrifuge tubes.** Disposable conical centrifuge tubes are used for centrifugation of cells as well as convenient sterile containers for liquid volumes ranging from ~5–50 ml. *Materials:* Polypropylene, polyethylene, or polycarbonate. *Sterilization:* γ-radiation. *Disposal:* Biohazard waste.
- **Autoclave supplies.** These include biohazard bags, autoclave indicator tape, sterilization pouches, etc. Biohazard bags are typically made with heavy-duty polypropylene, and they are used for containing biohazardous materials or wastes. Items to be autoclaved should be affixed with a piece of autoclave indicator tape. When sterilization conditions are met in the autoclave, the white stripes on the tape will change to a stark black color. Non-self-enclosed items such as scissors, spatula, etc. should be sealed in sterilization pouches. *Disposal:* Regular trash (after autoclaving).
- **Sterilization filters.** Liquids that contain heat-labile molecules such as proteins and drugs cannot be autoclaved, but they can be sterilized through sterile filtration. The pore size of sterile filters is typically 0.22 μm (or 0.1 μm for more stringent requirements), a size that does not allow passage of microbes. Small volumes of liquid (typically less than 50 ml) can be sterile filtered with syringe-tip filters, and larger volumes are typically handled with bottle-top filter units connected to vacuum. *Materials:* Various plastic materials. *Sterilization:* γ-radiation. *Disposal:* Biohazard waste.
- **Microcentrifuge tubes and vials.** Vials such as 0.5-ml or 1.5-ml microcentrifuge tubes are often used for liquid storage. For storage of cells in liquid nitrogen, a special kind of vial, the cryogenic vial, which is resistant to the extremely low temperature of –196°C, is required. The screw cap of a cryogenic vial typically has an O-ring that seals the content. *Materials:* Commonly polypropylene. *Sterilization:* γ-radiation or autoclave. *Disposal:* Biohazard waste.

I.3 Waste Disposal

The following are some of the typical wastes found in cell culture labs. *Note:* Institutional guidelines must be followed for disposal of wastes.

- **Biohazard sharps waste.** Heavy-duty, puncture-proof plastic bins should be used to contain sharps such as Pasteur pipettes that are used for aspiration, needles, glass slides, coverslips, etc.
- **Biohazard waste.** Containers lined with autoclavable biohazard bags should be used for plastic disposable items such as tissue culture vessels, serological pipettes,

etc. Often biohazard wastes can be decontaminated by autoclaving, after which they can be disposed of into regular trash. (The treated waste is required to be over-bagged with a dark or non-transparent trash bag to avoid confusion. Check your institution's policy regarding this.)

- **Special wastes.** Used chemicals and reagents should be collected into designated containers. As a general rule, do not mix liquid wastes and solid wastes if possible. For example, if a bottle is designated for trypan blue liquid waste, do not discard pipette tips, paper tissues, etc. into the bottle.
- **Regular trash.** Items that have not come in contact with any biological reagents, such as the wrap for serological pipette, or tissue paper that is used for disinfection with 70% alcohol, can be discarded in regular trash.

II Exercises

II.1 Know Your Cell Culture Lab Equipment

Survey the equipment in your cell culture lab and record your findings in Table 4.1.

TABLE 4.1

A Survey of the Equipment in My Cell Culture Lab

Equipment	Make	Model	Notes
Biosafety cabinet			[a]Class: ______
Aspiration pump			Vacuum: ______
Pipet-aid			
Micropipettes			Volumes: ______ ______ ______ Autoclavable/semi-autoclavable?
CO_2 Incubator			Settings Temperature: ______ CO_2%: ______
Autoclave			Suitable for liquid?
Freezer			Temperatures: ______
Cryo-storage			Tank volume: ______ Can-and-cane or rack-and-box?
Centrifuge			Centrifugal force: ______ Volumes: ______
Water bath			Settings Temperature: ______
Inverted microscope			[b]Objectives available: ______ [c]Color filters: ______ Phase contrast available?
Microscope camera			USB camera? Resolution: ______

[a] Note whether your BSC is a class I or II. (Class III is not typically used in cell culture labs.)
[b] Note the magnifications of the objectives such as 4x, 20x, etc.
[c] Note the colors such as red, blue, etc. of the color filters that are available.

II.2 Waste Disposals in Your Cell Culture Lab

Locate the waste disposal containers in your cell culture lab.

II.3 Liquid Handling

Working inside a BSC can be restricting and awkward. The following practices allow you to get used to working within a confined space using aseptic techniques before working in a BSC. You will practice using a pipet-aid and serological pipettes to transfer and dispense liquid; you will also be working with different tissue culture vessels.

Procedures

1. *Check-in*
   a. Reagents
      - A clear bottle of red-colored liquid (the "red buffer")
      - A nontransparent bottle with liquid inside
   b. Equipment and supplies
      - A T-25 tissue culture flask. *Note:* The neck of the flask is canted, and it can be positioned upright or laid down flat on one of its wide sides; cells will grow on the inner surface of this side.
      - A 15-ml conical centrifuge tube in a tube holder
      - A 6-well tissue culture plate
      - A pipet-aid
      - A 5-ml serological pipette
2. *Liquid transfer:* Practice the following steps to transfer 3.4 ml of the red buffer to a T-25 tissue culture flask:
   a. Sit down and make sure that your posture is comfortable and not awkward. Place the red buffer and the T-25 flask in upright position within easy reach. Loosen the caps of both.
   b. Read the graduation on the 5-ml serological pipette; most pipettes have two sets of graduation: one counts up from the tip for "liquid in" and the other counts down for "liquid out." Insert the pipette in the pipet-aid and turn the graduation reading for "liquid in" toward you so that you can read it easily.
   c. *Note:* Do not let the serological pipette touch anything except the inside of the containers. (Just imagine that every exposed surface is teeming with contaminants.) Also, try *not* to put your elbows down; this will help to reduce the risk of contamination when working in a real BSC.
   d. Right now you are holding the pipet-aid in one hand and your other hand is free. Now take off the cap of the buffer bottle with your free hand and grasp the outside of the cap with your fingers; do not touch the inside of the cap, and keep the opening of the cap facing down. Insert the serological pipette into the buffer bottle and draw 3.4 ml of the red buffer. Replace the cap of the red

buffer bottle. (Do not tighten it at this point.) Again using your free hand, first take off the cap of the T-25 flask with your thumb and index fingers and hold it; now grasp the flask with the rest of the figures. (Too hard? You can put the cap down on the bench facing *upwards* and hold the flask with your whole free hand, but this is less preferred.) Tilt the flask slightly to insert the serological pipette to near the bottom of the flask, and deliver the red buffer against one of its narrow sides. Remove the pipette, cap the flask, and lay the flask flat on the bench surface. Now tighten the caps of the bottle and the flask.

3. *Using the negative volume:* Again read the graduation on the 5-ml serological pipette; there is more above the line for "5 ml," and that is called the negative volume of the pipette. The negative volume can be added to the regular volume to transfer volumes that are >5 ml if necessary. It also serves as a "loading zone"; for example, you need to transfer 5.0 ml of liquid from a deep bottle, but while the pipette is inside the bottle, you cannot read the graduation. What you can do is to draw the liquid into the negative volume; now the pipette holds >5 ml of the liquid. Next, lift the pipette up so that you can read the graduation, and pipette the liquid out until the level drops down to the 5.0 ml mark. Practice the following steps using the negative volume and the techniques in Step 2:
   a. Place the nontransparent bottle and the 15-ml conical centrifuge tube in a tube holder in front of you. Loosen both caps.
   b. Using the negative volume of the pipette, draw 5.5 ml of liquid from the bottle. (Here you are using the negative volume to pipette >5.0 ml using a 5-ml pipette.) Add the liquid to the 15-ml tube. Tighten the caps.
4. *Dispensing multiple aliquots:* When transferring liquid to multiple receptacles, it is often easier to pipette in the liquid once and then dispense it into the receptacles in multiple aliquots. In the following exercise, you will pipette in 6.0 ml of the red buffer and dispense 1.5 ml into each of four wells in the 6-well plate:
   a. Place the red buffer bottle and the 6-well plate in front of you. Loosen the cap of the bottle.
   b. Pipette 6.0 ml of liquid from the bottle. (Again, use the negative volume.)
   c. Lift the lid of the 6-well plate; you can hold the lid in your hand with it facing down (preferred), or place it on the bench with it *facing up*. Lightly rest the tip of the serological pipette against the wall of the first well, and deliver 1.5 ml of the red buffer. Repeat this three more times. Try to control the volume as accurately as possible, but do not take too much time; each aliquot should take no more than a few seconds when you are working with cell cultures.
   d. Replace the lid of the plate, and tighten the cap of the bottle.

II.4 Basic Microscope Operation

When working with cell cultures, it is crucial to observe and monitor living cells; an inverted phase contrast microscope is used for such purpose. Detailed operations for microscopes of different makes may vary, but the underlying principles are similar. Use the following exercises to explore the features of your phase contrast microscope. Pay close attention to your instructor's demonstrations.

Procedures

1. *Check-in*
   a. Samples
      - Cell cultures in a 6-well plate
   b. Equipment
      - Inverted microscope equipped with phase contrast and camera
      - Adaptor for 6-well plate (if necessary)
2. *Power and light source:* Find the power switch of your microscope. The light source is usually a halogen lamp. Learn to adjust the light intensity.
3. *The mechanical stage:* If your microscope if equipped with a mechanical stage, then you can use the control rod to move the sample laterally on the stage with smooth, fine movements. Adaptor plates are used to hold different tissue culture vessels on the mechanical stage.
   a. Place a 6-well plate adaptor (if necessary) on the mechanical stage. Soak a piece of Kimwipe tissue paper with 70% ethanol and wipe the stage and the adaptor. ***Caution:*** Do not spray the microscope with ethanol (or any liquid for that matter).
   b. Ask your instructor for a 6-well plate with cell cultures. Place it on the adaptor. Move the stage along the x and y directions using the drop-down coaxial control rod. Use smooth movement so that the culture medium inside the plate wells does not slosh around.
4. *Observing through the eyepieces:* Even though images of the cells (or other specimens) can be captured on computer, it is still very important to observe them using your own eyes through the eyepieces of the microscope. *Practice:* Look through the eyepieces with *both* eyes (it's OK to keep your glasses on), and adjust the distance between the two eyepieces so that your view is one circle instead of two overlapping ones.
5. *Adjusting the focus:* Use the coarse and fine focus adjustment knobs on the two sides of the microscope to bring an image into focus. If the image is way out of focus, use lower magnification (see Step 6) or move an edge or other recognizable feature of the specimen into the view. Subsequently, use the coarse adjustment knob to bring the image into focus and use the fine adjustment knob to fine-tune it. *Practice:* Now adjust the focus until you see a clear image of the cells in the 6-well plate. Check with your instructor to make sure that you are indeed seeing the cells.
6. *Changing the magnification and adjusting the phase slider:* To change magnification, rotate the nosepiece and click the appropriate objective lens into place. ***Caution:*** Do not touch the objective lenses. For a phase contrast microscope, the phase annulus needs to be changed along with the magnification, which is done by moving the phase slider. Typically, a phase contrast slider has one slot for the 4x annulus, one slot for the 10x, 20x, and 40x annulus, and one open slot for brightfield (see Step 8). In addition, you may need to readjust the focus and the light intensity after changing the objective. *Practice:* Observe the cells with phase contrast at all the magnifications available for your microscope.

7. *Image capture:* For a *trinocular* microscope, which is equipped with a phototube for the camera, a beamsplitter splits the light emitted from the specimen between the eyepieces and the camera: It can either send all the light to the eyepieces or send a percentage of the light (between 50% and 100%, depending on the type of the microscope) to the camera. *Note:* The focus for the camera may be different from that for the eyepieces, so be sure to adjust the focus before capturing images. *Practice:* Take at least two phase contrast images of the cells using different magnifications or different views.
8. *Brightfield microscopy and the color filters:* When observing stained specimens, it is often best to use brightfield microscopy to observe the images in color. Some microscopes are equipped with color filters that can provide color balance or enhancement for the specimens. *Practice:* Learn how to switch from phase contrast to brightfield using the phase slider, and how to change the color filters if they are available.

II.5 Preparing the BSC

When not in use, the sash of the BSC is closed and a germicidal UV lamp is turned on. Before working in it with cell cultures, the BSC needs to be properly vented and disinfected to minimize the risk of contamination.

Procedures

1. *Turning on the BSC:* Turn off the UV lamp and turn on the light inside the cabinet. Lift the sash up. ***Caution:*** For some BSC models, if the sash is lifted too high or pulled too low, a loud alarm might sound. Turn on the blower—i.e., the pumps that direct airflow in the cabinet—and wait for at least 30 minutes. Particulates that may have accumulated inside the cabinet will be removed by the HEPA filter. (These steps may be performed by your instructor ahead of time.) *Your turn:* Find out where the switches are for the UV lamp, the light, and the blower. Find out the upper limit and the lower limit of the sash.
2. *Wiping down the BSC:* Tidy up loose sleeves and remove bracelet (if any). Put on latex gloves and spray both gloved hands with 70% ethanol, and then rub the two hands together to evenly spread the alcohol. (The gloves should be visibly wet.) Take a large tissue and tug it to roughly the size of a spread palm, and then spray it with 70% ethanol until it is soaked. Sit down in front of the biosafety cabinet. Press the soaked tissue with your spread palm and wipe down the bench surface of the BSC, reaching as far as you can. You can keep the alcohol-soaked tissue around for later use.

II.6 Preparing Cell Culture Medium

As previously mentioned, cell culture media typically contain salts, amino acids, vitamins, carbohydrates, growth factors, antibiotics, etc. In practice, culture media are often made in lab by mixing a medium "base" with supplements, and the result is called "complete medium" or "supplemented medium." Medium base contains mostly nonperishable (or not-so-perishable) small molecules such as salts, amino acids, and vitamins, and is commercially available in many varieties for use with different cells and different culture

Why Is 70% Ethanol Used as a Disinfectant But Not Pure Ethanol?

Ethanol can be purchased with 100% (200 proof) or 95% (190 proof) purity. However, in cell culture labs, ethanol is diluted with water to 70% by volume to be used as a disinfectant. The reason for using 70% ethanol is its ability to kill microbes such as bacteria and fungi, whereas 100% or 95% ethanol is actually less effective in doing so. Upon contact with 100% ethanol, proteins will quickly denature and aggregate in the microbial plasma membrane, which helps to block further permeation of ethanol into the cytosol, thus allowing the cells a chance to recover after the ethanol evaporates. With lower concentrations of ethanol, the denaturation and aggregation of plasma membrane proteins takes a longer time, thus allowing permeation of ethanol into the cytosol of the cells and subsequently denaturing most of the proteins, and killing the microbes for good. In addition, the slower evaporation of diluted ethanol allows longer contact time between ethanol and the microbes, which enhances its killing efficiency. It has been found that the optimal ethanol concentration for efficient killing of microbes is around 70%. As the ethanol concentration becomes higher or lower, it becomes less effective. For example, it is known that some microbes can survive up to 60% ethanol. In summary, 70% ethanol is used because it has the highest germicidal efficiency. The presence of a small amount of methanol does not affect the killing efficiency of ethanol. Isopropyl alcohol has similar germicidal potential as ethanol, and it is also used at a concentration of 70% by volume. In fact, 70% isopropyl alcohol (or isopropanol) is sold as rubbing alcohol, a topical first-aid antiseptic, in pharmacies. (On the other hand, isopropyl alcohol is too expensive for use as general disinfectant in lab, it takes longer to evaporate, and its smell is less pleasant than ethanol to some.)

conditions. These medium bases can be stored at 4°C for 1 year or longer. Supplements that are added to medium base typically include serum, which contains growth factors that promote cell proliferation, and antibiotics, which prevent microbial contamination. Sodium pyruvate, a metabolite, is added sometimes. Sera and antibiotics are perishable and must be frozen for long-term storage. Complete medium should be stored at 4°C and is generally good for 1–2 weeks. In the following exercise, your will use a medium base called minimum essential medium (MEM), and supplements to make complete MEM for later use.

Procedures

1. *Check-in*
   a. Reagents
      - [BSC] MEM (with L-glutamine and sodium pyruvate)
      - [BSC] Fetal bovine serum (FBS)
      - [BSC] 100X "Pen/strep" (penicillin and streptomycin mixture) stock solution
   b. Special equipment and supplies*
      - [BSC] Media bottle, 100 ml, sterilized by autoclaving
      - Labeling tape and Sharpie permanent marker
2. *Aseptic techniques*
   a. *Note:* All items must be thoroughly disinfected by 70% ethanol before being moved into the BSC, with the exception that pre-packaged sterilized items such as individually wrapped serological pipettes are handled differentially. Before you start, remember that: (1) gloves and tissues fresh from the box are *not* sterile; (2) your sterilized gloves become contaminated if you touch a non-sterile surface or object; and (3) the same goes for any other sterilized items.
   b. Spray and rub your gloves with 70% ethanol. Pick up the 100-ml media bottle and spray its surfaces with 70% ethanol; be sure to spray into the neck of the

* Specific items that are needed in addition to the consumable items list in Section I.2.

TABLE 4.2

Making Complete MEM with 10% FBS

Reagents	Volume (Pipetting Details)
MEM	89 ml *(use a 25-ml serological pipette with its negative volume for three transfers: 30 ml + 30 ml + 29 ml)*
Fetal bovine serum	10.0 ml *(use a 10-ml serological pipette)*
100X Pen/strep stock	1.0 ml *(use a 1-ml serological pipette)*

bottle. Next, wipe the surfaces of the bottle with a piece of ethanol-soaked tissue paper to spread the ethanol evenly. Move the bottle into the BSC (the ethanol will evaporate quickly). Use the same aseptic techniques to move the items labeled with [BSC] in Step 1 into the BSC.

3. *Making complete MEM.* For the following work, you will use the pipetting skills that you have practiced in Section II.3. (Except that now you will be working inside a real BSC. Relax. You'll be fine.) Again, make sure that you are sitting comfortably and that your arms can move freely. Do not put your elbows down on the air intake grill if you can help it; you would be blocking the airflow and increasing the risk of contamination.
   a. Loosen the caps of the reagent containers and the 100-ml media bottle, but do not take them off.
   b. Pipette reagents into the 100-ml media bottle according to Table 4.2. Make sure that the serological pipette touches the *inside* of sterile bottles or tubes. When finished, cap the media bottle and *swirl* the medium to mix. Do not shake.
   c. Label the bottle with "complete MEM + 10% FBS" on a piece of labeling tape with the permanent marker; do not write directly on the bottle. (Tip: Black Sharpie marker is the most resistant to 70% ethanol.) Also write down your group's name and today's date on the label.
   d. Remove the reagents from the BSC and store them properly. (You can leave the complete MEM in the BSC if you continue with Section II.7.)

II.7 Starting a Cell Culture from Cryo-Frozen Stock

In this session, we will start a cell culture of a mouse fibroblast cell line called L-929. For long-term storage, the cells are stored in liquid nitrogen in cryogenic vials. The cells must be first thawed, and then counted and plated in tissue culture vessels. The cell cultures will be used in Module III later.

Procedures

1. *Check-in*
   a. Reagents
      - [BSC] Complete MEM with 10% FBS (from Section II.6)
   b. Special equipment and supplies
      - [BSC] One T-25 tissue culture flask, sterile
      - [BSC] One 15-ml conical centrifuge tube, sterile
      - [BSC] Pasteur pipettes in a canister, autoclaved

- 1.5-ml microcentrifuge tubes
- Ice

2. *Preparation:* (Vent and wipe down the BSC if it has not been prepared already.) Move the items labeled with [BSC] in Step 1 into the BSC; remember to use the aseptic techniques that you practiced in Section II.6.
3. *Thawing and plating cells*
   a. *Note:* The L-929 cells will come to life once they are thawed, and because these cells are fibroblasts, they are prone to adhering to surfaces; therefore, the following steps need to be carried out as quickly as you can. But do not panic; you should have more than enough time.
   b. With your instructor's help, take a vial of frozen L-929 cells from the liquid nitrogen storage. Quickly thaw the frozen culture in a 37°C water bath. Use circular motion to swirl the vial, but keep the neck of the vial above the water level to avoid contamination. Immediately after the ice inside has all melted, stake the vial on ice and let it cool for a few seconds.
   c. Spray and wipe the vial with 70% ethanol and move it into the BSC. Loosen the caps of the vial and the 15-ml conical centrifuge tube. Using a 1-ml serological pipette, gently pipette up and down a few times inside the vial to resuspend the cells, and then transfer the cell suspension into the 15-ml conical centrifuge tube.
   d. Using a 10-ml serological pipette, add 9 ml of complete MEM to the cell suspension in the 15-ml tube; hold the pipette tip against the wall of the tube so that the liquid slides slowly. Gently pipette up and down a few times to mix, and then cap the tube.
   e. Place the 15-ml tube and a balance tube in a clinical centrifuge and run it at 1000 rpm for 3 minutes. ("Eyeball" balancing the two tubes is OK.) When finished, look at the bottom of the tube, and you should be able to see a small white pellet—those are the cells. Aseptically move the tube back into the BSC.
   f. Loosen the cap of the 15-ml tube. Open the lid of the canister that holds the Pasteur pipettes; hold the canister horizontally and shake the pipettes out bit by bit, and you should be able to pick one that is the farthest out without touching the others. Connect the wide end of the Pasteur pipette to the vacuum hose that is connected to the aspirator pump. Turn on the vacuum, and use the tip of the Pasteur pipette to aspirate the supernatant in the 15-ml tube, but be very careful to keep the pipette tip well away from the cell pellet or the pellet will be suctioned away. (You do not have to remove *all* the supernatant.) You have now successfully removed dimethylsulfoxide (DMSO), a cryo-protectant that helps the cells to survive the freeze-thaw process but is harmful to living cells.
   g. Add 5 ml of complete MEM to the 15-ml tube and gently pipette up and down to resuspend the cells; make sure that there is no visible clump. Pipette ~100 µl of the cell suspension into a 1.5-ml microcentrifuge tube, which is to be used for cell counting; cap the microcentrifuge tube and stake it on ice. (Note that the 1.5-ml tube is not sterile. Do not bring it into the BSC; bring the micropipette out instead. The pipette tip will be discarded after this. The cells used for counting will not be cultured further, so it's OK if they are not kept aseptic.)

h. *Note:* In normal lab procedures, you will count the cells at this point so that you can control the density of the cell culture. But for this exercise only, you will plate the cells first, so that you are not pressed for time when you learn how to count cells.

i. Place the T-25 flask upright. Loosen the cap, and pipette in 3 ml of complete MEM. Using the same pipette, transfer the cell suspension in the 15-ml tube to the T-25 flask. (The total volume in the flask is now ~8 ml.) Cap the flask, lay it down on the flat side, and gently rock the medium back and forth across the width of the flask to evenly spread the cells. *Important:* Do not allow the medium to reach the neck of the flask.

j. Label the flask with the name of the cell line (L-929), your group's name, and today's date. *Note:* You should write directly on the side of the flask instead of using a piece of labeling tape since the flask is disposable. Move the flask into a 37°C CO_2 incubator.

3. *Finishing up:* Discard the cryogenic vial and the 15-ml tube into the biohazard waste. Discard the Pasteur pipette into the biohazard sharps waste. Store the complete MEM at 4°C. Wipe down the biosafety cabinet with 70% ethanol.

II.8 Cell Counting

We will count cells with a hemacytometer in the following practices. The principle of cell counting will become clear as we carry out the procedures. Cell counting is done outside the BSC: While it is not necessary to keep the sample aseptic since these cells will not be cultured further, it is important to keep the sample from contaminating the environment. During cell counting, we will also determine the percentage of viable cells in the whole population by using a live stain, trypan blue. This blue dye is actively excluded by living, healthy cells, which remain colorless as a result, but not by unhealthy or dead cells, which will be stained blue.

Caution: Trypan blue is carcinogenic and can be absorbed through skin. Always wear gloves when working with trypan blue. Trypan blue waste should be collected in a special waste container.

Procedures

1. *Check-in*
   a. Reagents
      - Trypan blue
   b. Special equipment and supplies
      - Hemacytometer
      - (Optional) Compressed air (in compressed gas cylinder or duster can)
      - Hand tally counter
      - Container for trypan blue liquid waste
2. *Mixing cells with trypan blue:* Pipette 20 µl of the cell suspension (from Section II.7, Step 3g) to a 0.5-ml microcentrifuge tube. Be sure to pipette up and down to evenly suspend the cells before transfering them, because the cells have likely sedimented to the bottom. Add 20 µl of trypan blue solution to the 0.5-ml tube and gently pipette

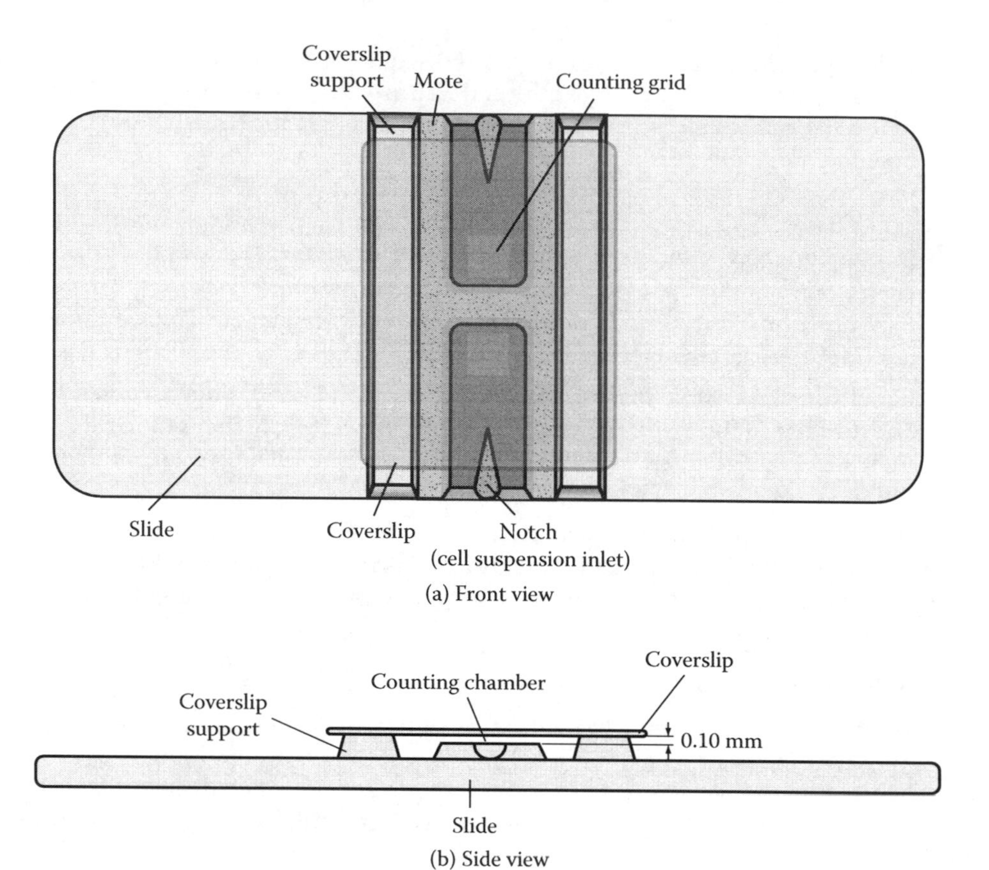

FIGURE 4.3
A hemacytometer is a precision instrument, the critical features of which are two precision-etched counting grids (see Figure 4.4) and a precise distance between the coverslip and the counting grids. The space between the coverslip and the etched surface below forms a counting chamber, and the volumes for the parts of the counting chamber that are marked by the grids can be precisely determined. When a cell suspension is injected into the counting chamber, the number of cells on the grids can be counted under a microscope, and since the volume is known, the cell count of the suspension (the number of cells per ml) can be determined. (a) Front view. (b) Side view.

up and down to mix it with the cells. Allow the cells to incubate with trypan blue for about 5 minutes. (You can move on to Step 3.)

3. *Preparing the hemacytometer.* A hemacytometer includes a slide with counting grids and a coverslip (Figure 4.3). Wash the slide and the coverslip with 70% ethanol, and then dry it with lint-free tissue paper or blow-dry it with the compressed air duster. Moisturize the coverslip supports and place the coverslip on top. (The small amount of water will act as adhesive to hold the coverslip down.)
4. *Loading the hemacytometer:* When the trypan blue/cell mixture is ready, pipette 20 μl, and then place the pipette tip on one of the notches of the hemacytometer (Figure 4.3). Gently pipette out the mixture, which will flow into the chamber counting by capillary action; inject enough mixture to cover the counting surface, but do not allow it to overflow. Load the counting chamber on the other side with the trypan blue/cell mixture as well.

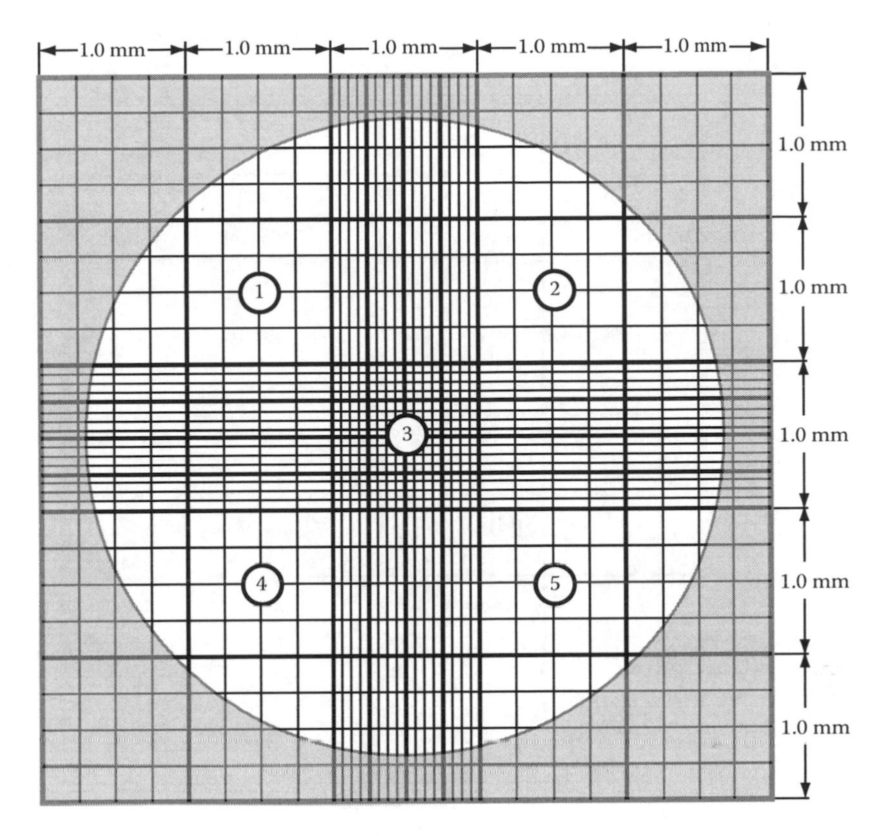

FIGURE 4.4
Etched grids on a hemacytometer. When viewed under the microscope at 40× magnification (lit circle), nine 1 mm² squares can be seen. Squares marked with numbers 1–5 are used for cell counting.

5. *Cell counting:* Place the hemacytometer under the microscope. Set the magnification of the microscope to 40x (using the 4x objective). Move one of the counting chambers into view, and adjust the focus so that a pattern like that in Figure 4.4 is clear and sharp. With a hand tally counter, record the cell counts in the four corner squares and the center square. Be sure to count the cells that are touching the middle lines on the top and the left of the square, and not the cells that are touching the middle lines at the bottom and the right of the square (Figure 4.5). Switch to the counting chamber on the other side of the hemacytometer and count five more squares. An ideal number of the total cell count would be between 200 and 400: if the cell count is too low, then the number may not be statistically significant; if the cell count is too high, then it may be hard to count individual cells, and the cell count may not be accurate.
6. *Calculating the cell count:* The total volume of the liquid marked by the 10 squares is 1.0 µl. Taking into consideration the twofold dilution by the trypan blue solution (Step 2), the cell count per ml is

$$\text{C.C.}\left(\text{ml}^{-1}\right) = N \times 10^3 \times 2 \qquad \text{(BLS II.1)}$$

where C.C. is the cell count, and N is the total number of cells in the 10 squares in Step 5.

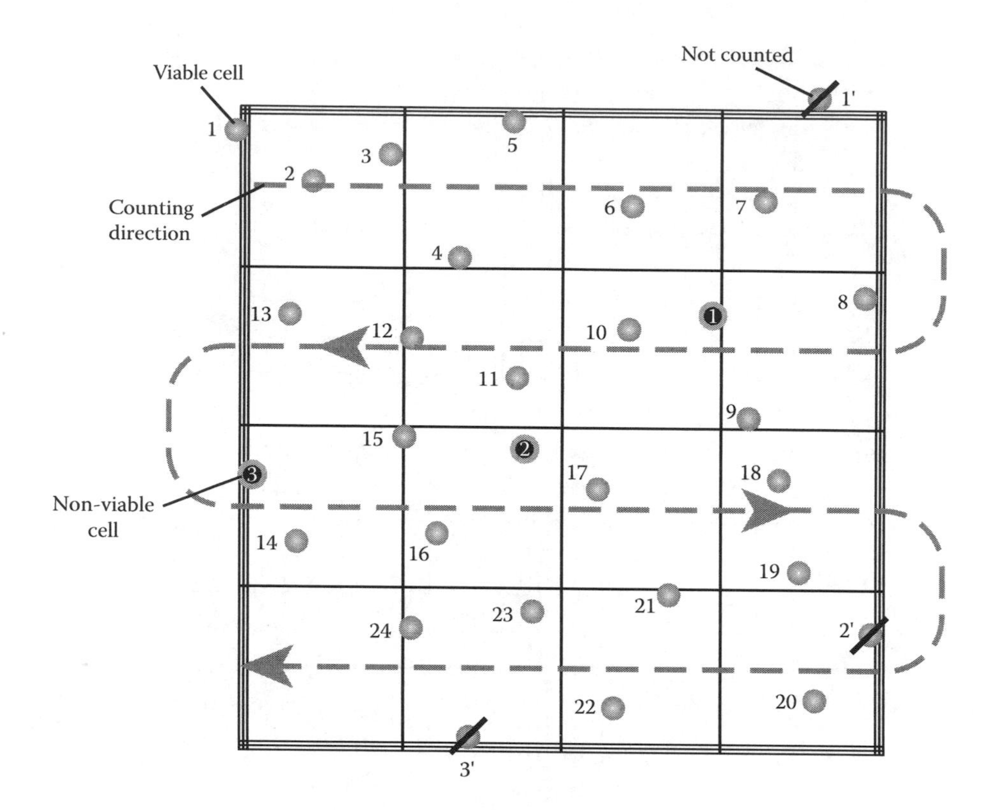

FIGURE 4.5
Counting cells. When using a hemacytometer with Neubauer ruling, the middle lines of the groups of three are the boundary lines for the 1 mm^2 squares. For each 1 mm^2 square, count cells totally within the square, and cells that touch the middle lines on the left (cell #1) and the top (cell #5) of the square, and not the cells that touch the middle lines on the right (cell 2′) and the bottom (cell 3′) of the square. The counting should start from the upper right corner, and "snake" back and forth until the lower right corner is reached. To determine the percentage of viable cells, count the number of nonviable cells that are darkly stained (cells #❶, ❷ and ❸) (switch to bright-field to visualize the staining if necessary), and the percentage of viable cells (89%) is the number of viable cells (24) divided by the total number of cells (27).

7. *Finishing up:* Pick up the coverslip with a pair of forceps; hold it over a trypan blue liquid waste container and spray-wash it with 70% ethanol. Spray-wash the chamber slide as well. After removing the trypan blue, use a tissue to gently rub mild detergent on the slide and the coverslip to remove any cellular residue, and then rinse them with water followed by de-ionized water. Place the hemacytometer on tissue paper to dry. Rinse the 0.5-ml microcentrifuge tube with 70% ethanol to remove residual trypan blue (the rinse goes into the trypan blue waste, too), and discard it into the biohazard waste along with the 1.5-ml microcentrifuge tube that contains the cell suspension.
8. *Homework:* Calculate the number of cells that you have seeded in the T-25 flask in Section II.7.

II.9 Changing Medium

Note: The following exercises can be performed several days after the previous exercises or on the same day using a cell culture that has been prepared by the instructor.

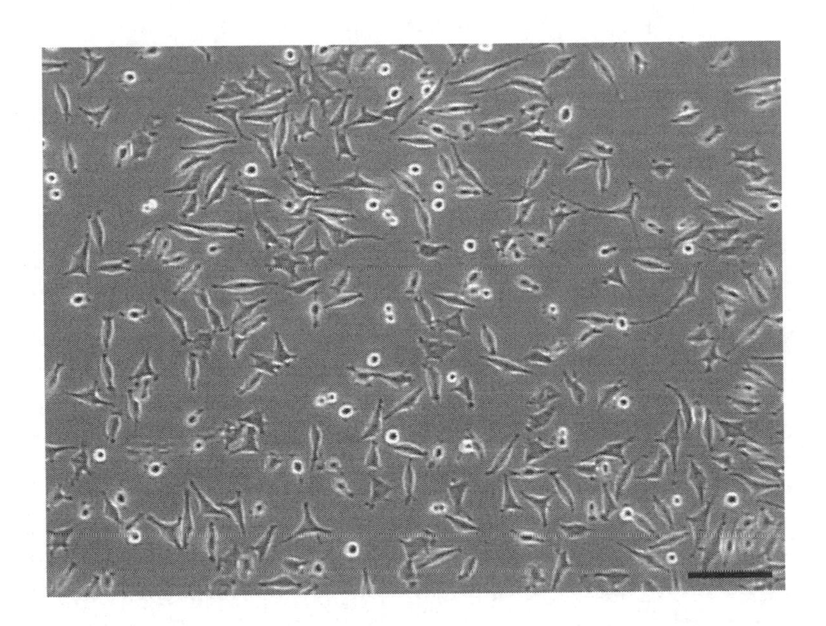

FIGURE 4.6
L-929 cells. Note that these cells tend to be spindle shaped, but overall the cell morphology tends to be quite heterogeneous. Another characteristic of L-929 cells is that they do not have extensive contacts with each other. Magnification: 100. Scale bar: 100 μm. Image is obtained using an inverted optical microscope set to phase contrast.

The cell culture that you have started in previous exercises is now proliferating. As the cells metabolize, the nutrients in the medium are gradually depleted and wastes have accumulated. To maintain the health of the culture, the medium needs to be changed every 2 or 3 days.

Procedures

1. *Check in*
   a. Reagents
      Complete MEM, warmed up to 37°C
2. *Preparation*
   a. Vent and wipe down the biosafety cabinet with 70% ethanol.
   b. Warm up the medium.
   c. Check the color of culture medium for your cell culture. It should be red and not purple or yellow.
   d. Observe the cells under the microscope. Note the morphology (Figure 4.6), which is an indication of the viability, and the confluency of the L-929 cells.
   e. Aseptically move the complete MEM and the cell culture into the BSC. Loosen the caps of the media bottle and the T-25 flask.
3. *Removing old culture medium:* Use a Pasteur pipette to aspirate the old medium. Place the tip of the Pasteur pipette in a corner of the flask, and take care *not* to scratch the cells.

4. *Adding new medium:* Add 8 ml of new medium to the flask. Do *not* splash the medium on the cell growth surface or the cells will be dislodged.
5. *Finishing up:* Label the flask on the side with the date of medium change and the initials of the person who performed the change. Move the flask back to the 37°C CO_2 incubator. Store the complete MEM at 4°C.

II.10 Passaging and Freezing Cells

Most adherent cells grow as a monolayer in *in vitro* cultures. When the cells proliferate to the point where there is no more surface left, the culture has reached confluency. At this point, a "contact inhibition" mechanism will be triggered in the cells, and the cells will stop dividing and begin to die off in a matter of days. In addition, changes in gene regulation and expression often occur in confluent cells. For these reasons, a cell culture needs to be "split" before it reaches confluency to keep it healthy and to minimize mutation. Adherent cells must be detached from the growth surface before splitting, and this is achieved through the use of a protease, trypsin, that degrades cell surface proteins, including the adhesion proteins. Cells can thus be "lifted" off the growth surface and suspended. The cell suspension can then be diluted or divided before seeding a new culture. This process is called cell passaging.

Cells can be stored long term in liquid nitrogen, but they must be properly frozen first. To freeze adherent cells, they are first lifted off the growth surface and then suspended in a freezing medium at desired cell count, typically ~1×10^6 cells/ml. Freezing of the cells requires a slow and steady drop of the temperature; we will take advantage of the refrigerator and freezers available in the lab to achieve that goal. To protect the cells during the freeze-thaw cycle, dimethylsulfoxide (DMSO) is added to the freezing medium as a cryoprotectant to reduce the formation of ice crystals in the cells that can cause cell lysis.

Procedures for Cell Passaging

Note: In this exercise you will split the cells with a 1:5 ratio; that is, the cell density will be reduced by 80%.

1. *Check-in*
   a. Materials and samples
      - [BSC] Your L-929 cell culture from previous exercises in a T-25 flask, grown to 85–95% confluency
   b. Reagents
      - [BSC] Complete MEM, warmed up in 37°C water bath
      - [BSC] Trypsin ethylene diamine triacetic acid (EDTA), warmed up to room temperature
      - [BSC] PBS
   c. Special equipment and supplies
      - [BSC] Two T-25 flasks, sterile
2. *Preparation*
   a. Vent and wipe down the biosafety cabinet. Aseptically move the items labeled with [BSC] in Step 1 inside the BSC. Loosen all caps.
   b. Observe the cells under the microscope and note the morphology of the cells and the confluency of the culture.

3. *Trypsinization*
   a. Aspirate the medium from the cell culture in the T-25 flask. Pipette 5 ml of PBS into the flask, and gently slosh the PBS back and forth over the cells to rinse off residual medium. This step removes residual serum proteins that inhibit trypsin.
   b. Aspirate the PBS, and pipette 0.5 ml of trypsin EDTA solution onto the cells. Tilt or rock the flask so that the enzyme solution completely coats the cells. Place the flask in the 37°C CO_2 incubator for 1–2 minutes.
   c. Observe the cells under the microscope and note the shape changes. ***Caution:*** Do not allow the cells to be in contact with (uninhibited) trypsin for more than 5 minutes.
4. *Suspending the cells*
   a. Bring the flask aseptically back to the BSC. Tap the side of the flask several times to dislodge the cells. Add 4.5 ml of complete MEM to the flask. Using the same pipette, wash the medium over the cell-growing surface several times to dislodge the remaining cells. *Note:* The serum in the complete medium will inhibit trypsin.
   b. Pipette up and down several times more to disperse any cell clumps. To avoid foaming, lift the tip of the serological pipette out of the liquid when pipetting out, and direct the liquid against a surface. The medium should appear slightly turbid at this point because of the suspended cells that scatter light.
5. *Splitting the cells*
   a. Add 7 ml of complete MEM into each of the two new T-25 flasks (in upright position).
   b. Pipette the cell suspension in the old T-25 flask up and down twice more (in case the cells have sedimented), and then pipette 1 ml of the cell suspension each to the two new T-25 flasks. Pipette up and down a few times in the new medium to ensure full delivery of the cells. Now each new flask has 1/5 of the original cell population.
   c. Lay the new flasks down and gently rock the medium back and forth to disperse the cells evenly.
   d. Label the new flasks with the name of the cells, the split ratio, the passage number (P1, for example; ask your instructor for the previous passage number), the date, and your group's name. Place the new cultures in the 37°C CO_2 incubator.
6. *Finishing up:* Store the complete MEM and the trypsin. Discard the old T-25 flask with the leftover cells into the biohazard waste. Wipe down the biosafety cabinet with 70% alcohol.

Procedures for Freezing Cells

In this exercise you will freeze one 80–95% confluent cell culture from a T-25 flask; you will be using the skills you have learned in cell passaging and cell counting.

1. *Check-in*
   a. Materials and samples

- [BSC] One L-929 cell culture in T-25 flask, 80–95% confluent

b. Reagents
   - [BSC] Complete MEM, warmed up to 37°C
   - [BSC] Cell freezing medium, containing 10% DMSO, 10% FBS, and 80% complete MEM
   - [BSC] Trypsin EDTA, warmed up to room temperature

c. Special equipment and supplies
   - [BSC] One 15-ml conical centrifuge tube, sterile
   - [BSC] One 2.0-ml cryogenic vial, sterile
   - One 1.5-ml microcentrifuge tube
   - One 0.5-ml microcentrifuge tube
   - Cell-freezing container

2. *Preparation*
   a. Vent and wipe down the BSC with 70% ethanol. Aseptically move the items labeled with [BSC] in Step 1 into the BSC.
   b. Label the 15-ml centrifuge tube with your group's name, and the cryogenic vial with the name of the cells (L-929), your group's name, and today's date. Loosen all caps.
   c. As always, observe the cell culture under the microscope and note its morphology and confluency.

3. *Trypsinizing and suspending the cells:* The following are essentially the same procedures that were used for cell passaging in the previous exercise:
   a. Aspirate the old medium, and then rinse the cells with 5 ml of PBS.
   b. Aspirate the PBS, and then add 0.5 ml of trypsin EDTA solution to coat the cell surface. Trypsinize for 1–2 minutes at 37°C. Observe with the microscope to confirm that the cells are detached.
   c. Dislodge cells by tapping the flask. Add 4.5 ml of complete media to suspend the cells.
   d. Transfer ~100 µl of the cell suspension to a 1.5-ml microcentrifuge tube for cell counting.
   e. Transfer the rest of the cell suspension into the 15-ml conical centrifuge tube and place it on ice.

4. *Cell counting:* Count the cells in the 1.5-ml microcentrifuge tube using the same cell counting procedure in Exercise II.8. The following is an outline of the procedures:
   a. Mix 20 µl of the cell suspension with 20 µl of trypan blue.
   b. Load the hemacytometer with the cell/dye mixture.
   c. Count the number of cells in 10 squares total.
   d. Calculate the number of cell count per ml:

$$\text{C.C. } \left(\text{ml}^{-1}\right) = N \times 10^{3} \times 2$$

   with N as the total number of cells in 10 squares.

   e. Calculate the total cell count of the cell suspension from Step 3.
5. *Centrifuging the cells*
   a. Centrifuge the cell suspension from Step 3 at 1000 rpm for 3 minutes.
   b. Remove the supernatant; again, be very careful not to aspirate off the cell pellet.
   c. Using the total cell count obtained from Step 4e, calculate the volume of medium needed to make a suspension with a cell count of 1×10^6/ml.
6. *Resuspending the cells in freezing medium:* Add freezing medium with the volume calculated in Step 5c to the cell pellet. Pipette up and down until the cells are completely resuspended.
7. *Freezing the cells*
   a. Transfer 1 ml of the resuspended cells to the 2.0-ml cryogenic vial. Add the cell density (1×10^6/ml, for example) to the labeling on the vial.
   b. Place the cryogenic vial in a cell-freezing container. *Note:* A cell-freezing container uses isopropanol as a bath to help achieve a steady rate of cooling, which is crucial to cell freezing.
   c. [By the instructor] Cool the cell-freezing container in a 4°C refrigerator for 30 minutes, then transfer to a –80°C freezer. After ~24 hours, transfer the cryogenic vials to a liquid nitrogen tank for long-term storage.
8. *Finishing up:* Clean the hemacytometer, and properly discard the disposable items. (By now you should know how to dispose of them.) Store the reagents. Wipe down the biosafety cabinet with 70% alcohol.

Appendix. Recipes and Sources for Equipment, Reagents, and Supplies*

I.1 Equipment for Cell Culture

- **Biosafety cabinet.** Labconco Purifier Delta digital series safety cabinet, mfr. no. 3880204, Fisher cat. no. 16-107-209. *Sel. crit.:* Class IIA BSC that offers personnel, product, and environment protection; UV lamp recommended.
- **Aspiration pump.** Welch standard-duty dry vacuum pump, model 2522B-01, Fisher cat. no. 01-051-1A. *Sel. crit.:* Economic, oil-free dry vacuum pump (they work very well as aspiration pumps).
- **Pipet-aid.** Drummond portable pipet-aid filler/dispenser, multispeed XP, mfr. no. 4000101, Fisher cat. no. 13-681-15E. *Sel. crit.:* With adjustable speed; a rest stand is recommended.
- **Autoclave (pressurized steam sterilizer).** Napco sterilizer, model 8000-DSE, mfr. no. 51220025, Fisher cat. no. 14-487-2. *Sel. crit.:* Bench-top model is sufficient, best if equipped with slow exhaust, which allows autoclaving of liquid.

* *Disclaimer*: Commercial sources for reagents listed are used as examples only. The listing does not represent endorsement by the author. Similar or comparable reagents can be purchased from other commercial sources. See selection criteria (*sel. crit.*).

- **Liquid nitrogen storage.** Barnstead/Thermolyne Bio-Cane can-and-cane cryogenic system, Model 34, mfr. no. CK509X3, Fisher cat. no. 11-675-92. *Sel. crit.:* A can-and-cane type of cryo-storage is economic and convenient for small numbers of samples.
- **Centrifuge.** IEC Centra CL2 benchtop clinical centrifuge, mfr. no. 426, Fisher cat. no. 05-101-7; rotor, mfr. no. 236, Fisher cat. no. 05-101-9; tube inserts, mfr. no. 2092S, Fisher cat. no. 05-101-14. *Sel. crit.:* Should include a rotor that holds 15-ml conical centrifuge tubes.
- **Inverted microscope.** Olympus CKX41, Meiji TC5400, Motic AE31, etc. all have the necessary features for cell culture works. *Sel. crit.:* Inverted microscope with phase contrast and a phototube with trinocular configuration.
- **Image capture.** Moticam 1000. *Sel. crit.:* Digital camera with USB connection (USB camera preferred for portability and user-friendliness).

I.2 Consumable (Disposable) Items

- **Serological pipettes (various volumes).** *Sel. crit.:* Polystyrene pipettes, sterile, and individually wrapped with paper and plastic.
- **Pasteur pipette.** Fisherbrand disposable borosilicate glass Pasteur pipettes, Fisher cat. no. 13-678-20C. *Sel. crit.:* 9″ length preferred.
- **T-25 and T-75 tissue culture flasks.** BD Falcon Primaria tissue culture flasks, 25 cm^2, mfr. no. 353808, Fisher cat. no. 08-772-45. *Sel. crit.:* Vented cap, canted neck, tissue culture treated growth surface, and sterilized with γ-radiation.
- **15-ml and 50-ml conical centrifuge tubes.** (Available from many suppliers or in on-campus storerooms. Feel free to use the most economic ones available.) *Sel. crit.:* Made with polypropylene or polyethylene, graduated, and sterilized with γ-radiation.
- **0.5-ml and 1.5-ml microcentrifuge tubes.** (Available from many suppliers or in on-campus storerooms. Feel free to use the most economic ones available.) *Sel. crit.:* Made with polypropylene, graduated, with snap caps, and autoclavable.
- **Bottle-top sterile filter units (various sizes).** Millipore Steritop sterile vacuum bottle-top filters, 150 ml, Fisher cat. no. SCGPT01RE. *Sel. crit.:* Pore size 0.22 μm or less, neck size 45 mm preferred for interface with media bottles, low protein binding preferred, sterilized with γ-radiation, and individually wrapped.
- **Syringe-tip sterile filter unit.** Millipore Millex sterile syringe filter units, Fisher cat. no. SLGP033RS. *Sel. crit.:* Pore size 0.22 μm or less, low protein binding preferred, sterilized with γ-radiation, and individually wrapped.
- **Sterilization pouches (various sizes).** Fisherbrand instant sealing sterilization pouches, 3.5″ × 5.25″, Fisher cat. no. 01-812-50. *Sel. crit.:* Paper and plastic, self-sealing, and color indicator.
- **Autoclave indicator tape.** Fisher cat. no. 11-889-14. *Sel. crit.:* With color indicator.
- **6-well tissue culture plate.** BD Falcon 6-well tissue culture plates, mfr. no. 353046, Fisher cat. no. 08-772-1B. *Sel. crit.:* Tissue culture treated growth surfaces, sterilized with γ-radiation, and individually wrapped.

- **Media bottles (various volumes).** Pyrex reusable media bottles (100 ml), mfr. no. X1395100, Fisher cat. no. 06-414-1A. *Sel. crit.:* Screw thread size GL45, and autoclavable.

I.3 Waste Disposal

- **Biohazard sharps waste bins (various sizes).** Fisherbrand Sharps-A-Gator sharps containers, 1 gallon, Fisher cat. no. 14-827-101. *Sel. crit.:* Made with polypropylene, and puncture resistant. *Important:* Follow institutional guidelines.
- **Biohazard waste bags (various sizes).** Fisherbrand biohazard autoclave bag, 14″ × 19″, Fisher cat. no. 01-828B. *Sel. crit.:* Bags should be made with polypropylene. *Important:* Follow institutional guidelines.

II.3 Liquid Handling

- **The "red buffer."** Shake the tip of a red marker with *water-soluble* ink in a bottle of water until the color of water turns sufficiently red.

II.6 Preparing Cell Culture Medium

- **Minimum essential medium (MEM).** American Type Culture Collection (ATCC (Manassas, VA)), cat. no. 30-2003. *Sel. crit.:* EMEM should contain Earle's Balanced Salt Solution (EBSS), nonessential amino acids, 2 *mM* L-glutamine, 1 *mM* sodium pyruvate, and sodium bicarbonate.
- **Fetal bovine serum (FBS).** Thermo Scientific HyClone fetal bovine serum, Fisher cat. no. SH3062603. *Sel. crit.:* Sterile filtered, non-heat-deactivated, preferably with non-U.S. origin (to minimize the risk of BSE contamination).
- **Penicillin and streptomycin stock solution (Pen/strep).** Thermo Scientific HyClone penicillin and streptomycin mixture in 0.85% NaCl solution, Fisher cat. no. SV30010. *Sel. crit.:* 100X concentrated stock solution with 10,000 U/ml of penicillin and 10,000 μg/ml of streptomycin, in NaCl solution or buffered solution.

II.7 Starting Cell Cultures from Frozen Stock

- **L-929 cells.** ATCC, cat. no. CCL-1™. *Sel. crit.:* None.

II.8 Cell Counting

- **Trypan blue.** Sigma-Aldrich, cat. no. T8154. *Sel. crit.:* Concentration 0.4% in NaCl solution or buffered solution; sterility optional.
- **Hemacytometer.** Bright-Line counting chambers, mfr. no. 3100, Fisher cat. no. 02-671-10. *Sel. crit.:* Similar hematocytometer with Neubauer ruling.
- **Compressed air duster.** Staples X80004 compressed air duster (10 oz. can), part no. E02-MIS-X80004. *Sel. crit.:* Compressed air or N2 from cylinder will work, too.
- **Hand tally counter.** Fisher cat. no. 07-905-6. *Sel. crit.:* Similar function.

II.10 Passaging and Freezing Cells

- **Trypsin EDTA.** Lonza BioWhittaker Trypsin-Versene, Fisher cat. no. BW17-161E. *Sel. crit.:* Trypsin concentration 0.5 g/L (0.05%), EDTA (Versene) concentration 0.2 g/L (0.02%), and (preferably) tissue culture tested.
- **PBS.** Dulbecco's Phosphate Buffered Saline (D-PBS), Sigma-Aldrich, cat. no. D8537. *Sel. crit.:* Without Ca^{2+} and Mg^{2+}, tissue culture tested.
- **Cell freezing medium (10 ml).** Mix 8.0 ml of complete MEM, 1.0 ml FBS, and 1.0 ml of sterile-filtered DMSO (Sigma-Aldrich, cat. no. D2438).
- **Cryogenic vial.** Fisher cat. no. 05-669-64. *Sel. crit.:* Volume 1.8–2.0 ml, externally threaded, self-standing, and sterilized with γ-radiation.
- **Cell-freezing container.** Nalgene "Mr. Frosty" freezing container, mfr. no. 5100-0001, Fisher cat. no. 15-350-50. *Sel. crit.:* Similar configuration and function.

5

Module III. Biocompatibility Testing: Cytotoxicity and Adhesion

The concept of biocompatibility can be divided into principles: *biosafety* and *biofunctionality* (Kirkpatrick et al. 1998). The biosafety principle requires materials to "not, either directly or through the release of their material constituents: (i) produce adverse local or systemic effects, (ii) be carcinogenic, or (iii) produce adverse reproductive and developmental effects" (FDA 1995). On the other hand, biofunctionality concerns the performance of a given material in application to a particular tissue; in this regard, biofunctionality can be defined as "the ability of a material to perform with an appropriate host response in a specific application" (Williams 1987). An example that illustrates these two principles is the biocompatibility requirements for orthopedic implants: for the biosafety requirement, these devices need to be made of materials that are nontoxic; for the biofunctionality requirement, bone-forming cells must be able to adhere to the surface of the implant materials to promote bone growth.

In the early history of biomaterials, determination of biocompatibility was based on empirical observations. As these empirical observations accumulate, we can sometimes predict the biocompatibility of some materials, especially some of the inorganic ones such as some of the metals, and ceramics. Today, procedures for biocompatibility testing of materials in medical devices have been standardized by a number of organizations that include the American Society for Testing and Materials (ASTM), the International Organization for Standardization (ISO), and U.S. Pharmacopoeia (USP). These standards specify specimen requirements, testing conditions, controls, and data interpretation for a series of *in vitro* and *in vivo* tests. In the United States, the Food and Drug Administration (FDA), which regulates the marketing of medical devices, has provided guidance (FDA 1995) to selecting "appropriate tests to evaluate the adverse biological responses to medical devices." The approach to test selection for the "501(k)," or Pre-Marketing Notification, is summarized in Figure 5.1. If toxicological data are required for a particular medical device, then initial tests for biocompatibility evaluation should be conducted according to Table 5.1, and the results are included for a Pre-Market Approval (PMA) of the device. Some of these tests—the cytotoxicity test, for example—are designated to address the biosafety issue, which is naturally a major concern for medical device. On the other hand, different tests are required depending on the specifics of the application in terms of the types of contact (surface, or similar conditions), the contacted tissues (skin, blood, etc.), and the contact duration (<24 hours, 24 hours to 30 days, or >30 days, for example); these application-specific tests address the biofunctionality issue besides the biosafety issue.

Another issue in biocompatibility testing is the use of *in vitro* testing vs. *in vivo* testing. In *in vivo* testing, animal models are used to simulate the human physiological environment; therefore, *in vivo* testing provides more complete and reliable information, and is necessary for safe use of medical devices in humans at certain stages of the research and development process. On the other hand, *in vitro* testing is relatively fast, easy, and inexpensive, and it helps to minimize animal use. Moreover, with the vast knowledge in molecular and cell biology that we have today, *in vitro* testing can be used to understand the mechanism

Biocompatibility Testing for Breast Implants

In 2006, the FDA approved marketing of silicone-gel breast implants after a 14-year ban (http://www.fda.gov/cdrh/breastimplants/siliconegel.html). Controversies aside, let us take a look at the approval process and see how biocompatibility testing was conducted on silicone-gel breast implants to meet FDA requirements. According to Table 5.1, a breast implant is an implant device that comes in contact with tissues with a contact duration definitely longer than 30 days. The following tests were conducted by Mentor Corporation for its MemoryGel™ silicone gel-filled breast implants to obtain premarket approval by the FDA (PMA# P030053). The major components of the implants—that is, the shell (silicone elastomer), the patch (silicone elastomer), and the gel (cross-linked and platinum-cured silicone gel), along with the whole device—are among the test materials.

- Cytotoxicity test: Effects of the test materials on the viability of mouse fibroblast L-929 cell cultures were tested using the agar overlay method and the elution method. (*Note:* The agar overlay method is similar to the direct contact method except that a layer of soft agar is set in between the cell monolayer and the test material.)
- Sensitization test: The immune systems of guinea pigs were first challenged with injection of and skin exposure to extracts of the test materials; after a delayed period, the skin of these guinea pigs was exposed to the test materials again and watched for any redness or swelling.
- Short-term irritation (and implantation) test: Rabbits were subcutaneously injected with extracts of the test materials and watched for any redness and swelling of the skin. (*Note:* Irritation is a result of localized inflammatory response, rather than an immunological response such as sensitization.)
- Acute systemic toxicity test: Relatively large doses (50 ml per kg body weight) of extracts of the test materials were administered to mice through peritoneal injection. The mice were then observed for any sign of toxicity such as convulsion, prostration, weight loss, etc.
- Immunotoxicity test: Samples of the shell materials were implanted into mice for a prolonged period (10–28 days). The immune responses of these mice were monitored by measuring immunological parameters such as the white blood cell count, antibody production, etc.
- Genotoxicity test: Four different *in vitro* tests were conducted using the test materials or their extracts to determine whether the test materials can cause gene mutation, DNA damage, or chromosomal aberrations.
- Haemocompatibility test: This test was also done *in vitro*. Red cells from rabbit blood were incubated with the test materials or their extracts, after which the extent of hemolysis, i.e., the rupturing of red cells, was measured.
- Pyrogenicity test: Rabbits were intravenously injected with extracts of the test materials and then monitored for any fever response.
- Carcinogenicity test: The silicone gel was implanted into albino rats. The animals were then monitored for a long period (>24 months or lifetime) and observed for the occurrence of tumors and the nature of these tumors. (This test was not required by the FDA guidance since the results of their genotoxicity test were negative; it was conducted nevertheless. This test would be required if existing data suggest tumorigenic potential of the test materials.)
- Reproductive toxicity and teratogenicity test: The reproduction potential and birth defects were monitored for rats that were exposed to extracts of the test materials or were implanted with the test materials. (*Note:* Even though this test is not required according to Table 5.1, but considering the breast is an important organ for reproduction, a test of this nature is appropriate.)
- Subacute and subchronic toxicity tests were not conducted because chronic toxicity data are available for the test materials.

of cellular responses to materials at the molecular level, and such understanding can in turn help us to improve the biocompatibility of materials.

In this module, we will study both the biosafety and the biofunctionality aspects of biocompatibility testing using *in vitro* methods. As shown in Table 5.1, cytotoxicity, or toxicity to cells, is one of the first properties to be tested when determining the biosafety of a prospective material. In our experiments, we will test the cytotoxicity of several materials using *in vitro* cell cultures. For biofunctionality, we will study how well the test materials support adhesion by bone-forming cells which results in cell proliferation on the surfaces of the materials, as such studies help to illustrate the prospects of these materials as bone-repair or bone-replacement materials.

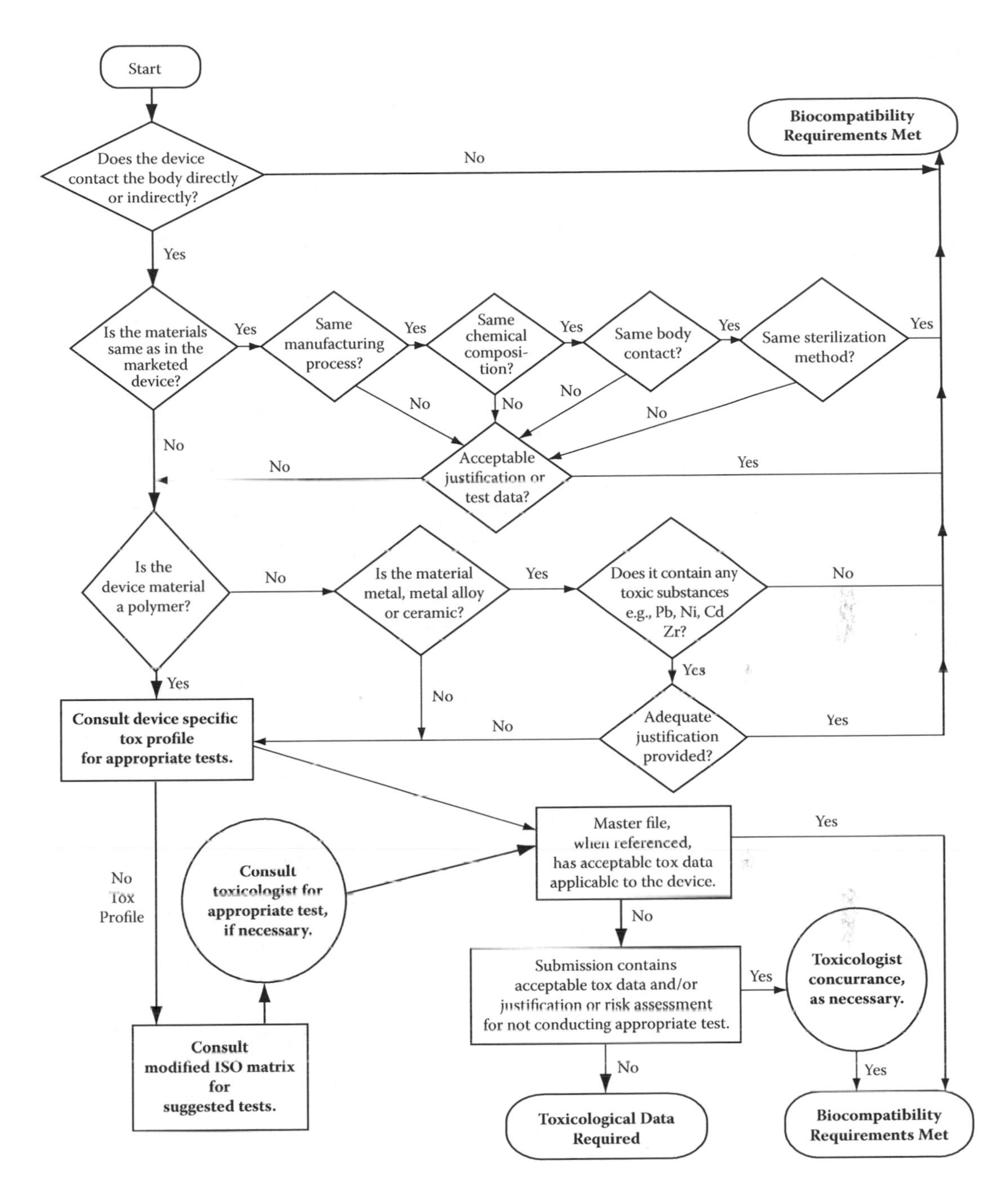

FIGURE 5.1
Biocompatibility flow chart for the selection of toxicity tests for 510(k) provided by the FDA (FDA 1995). "510(k)" is also known as Pre-market Notification, and it is a requirement that medical devices must be registered with the FDA before being marketed. As part of this requirement, manufacturers must decide whether toxicity tests need to be conducted for their devices according to this flow chart. For example, if a manufacturer intends to use a novel polymer as a breast implant material and no toxicological data are available for this material, then according to the criteria in this flow chart (bold arrows), toxicological tests must be conducted. In such cases, the devices will need to go through the Pre-market Approval process, in which toxicological data along with other data are reviewed by the FDA. Guidelines for obtaining relevant toxicological data are summarized in Table 5.1.

TABLE 5.1

Initial Tests for Biocompatibility Evaluation of Medical Devices (FDA 1995)

Device Categories		Biological Effect								
Body Contact		**Contact Duration** **A: Limited (24h)** **B: Prolonged (24h to 30 days)** **C: Permanent (>30days)**	**1. Cytotoxicity**	**2. Sensitization**	**3. Irritation or Intracutaneous Reactivities**	**4. System toxicity (acute)**	**5. Su-chronic toxicity (subacute toxicity)**	**6. Genotoxicity**	**7. Implantation**	**8. Haemocompatibility**
Surface devices	Skin	A	x	x	x	.	.	.	.	.
		B	x	x	x	.	.	.	.	.
		C	x	x	x	.	.	.	.	.
	Mucosal membrane	A	x	x	x	.	.	.	.	.
		B	x	x	x	o	o	.	o	.
		C	x	x	x	o	x	x	o	.
	Breached or compromised surfaces	A	x	x	x	o	.	.	.	.
		B	x	x	x	o	o	.	o	.
		C	x	x	x	o	x	x	o	.
External communicating devices	Blood path, indirect	A	x	x	x	x	.	.	.	x
		B	x	x	x	x	o	.	.	x
		C	x	x	o	x	x	x	o	x
	Tissue/bone/dentin communicating[a]	A	x	x	x	o	.	.	.	.
		B	x	x	o	o	o	x	x	.
		C	x	x	o	o	o	x	x	.
	Circulating blood	A	x	x	x	x	.	o[b]	.	x
		B	x	x	x	x	o	x	o	x
		C	x	x	x	x	x	x	o	x
Implant devices	Tissue/bone	A	x	x	x	o	.	.	.	.
		B	x	x	o	o	o	x	x	.
		C	x	x	o	o	o	x	x	.
	Blood	A	x	x	x	x	.	.	x	x
		B	x	x	x	x	o	x	x	x
		C	x	x	x	x	x	x	x	x

x = ISO Evaluation tests for consideration
o = Additional tests which may be applicable
[a] Tissue includes tissue fluids and subcutanous spaces
[b] For all devices used in extracorporeal circuits

Session 1. Cytotoxicity Evaluation Using Direct Contact Tests

Cytotoxicity refers to toxic effect at the cellular level, which can be due to DNA damage, permeabilization of the cellular membrane, apoptosis, and other causes. In cytotoxicity testing, cellular damages caused by the test material are mostly chemical in nature, and the harmful chemicals are often leachable chemical additives in the material or even degradation products of the material itself. Since cytotoxicity testing uses *in vitro* methods, is relatively easy and inexpensive to perform, and provides critical information on the cytocompatibility of the material, it is often used as a first step in determining the biosafety of a given material.

In Sessions 1 and 2, we will follow the standard procedures described in ISO 10993-5 (ISO 1999) to test the cytotoxicity of a number of materials. In Session 1, we will use the direct contact method. In this method, the material sample is placed in direct contact with cells in *in vitro* culture (Figure 5.2). The material sample is required to have a (more or less) flat surface with an area of ~100–250 mm^2, and after properly sterilized, it is placed

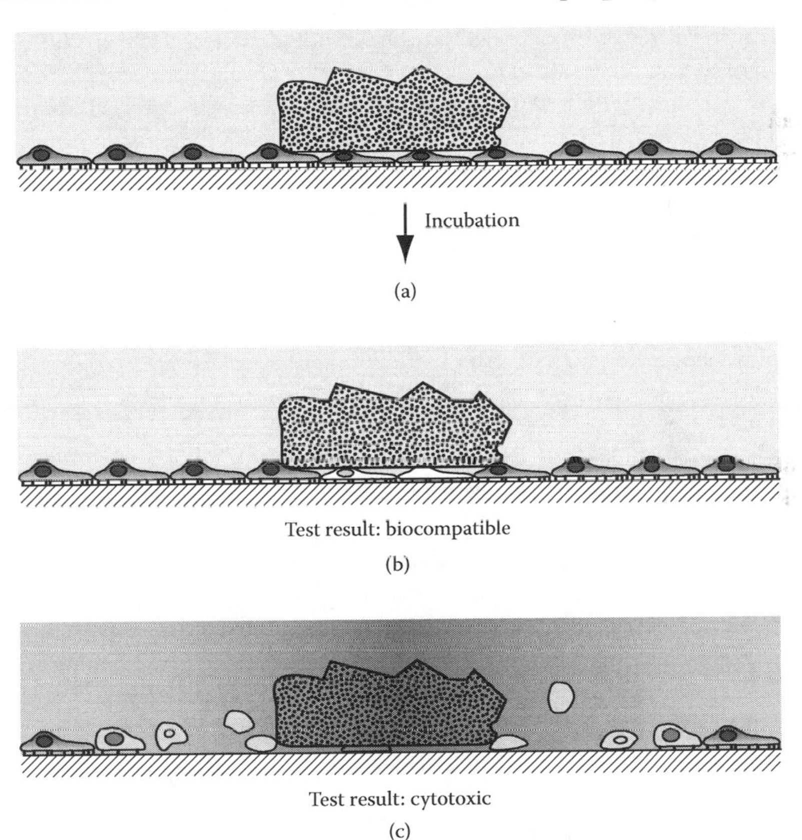

FIGURE 5.2
Cytotoxicity evaluation of material using the direct contact test. (a) A piece of test material is placed directly on a monolayer of cells and incubated for a period of time, after which the cells are observed for signs of cytotoxicity. (b) Materials that do not damage the cells are deemed noncytotoxic. Note that cells directly underneath the sample might be damaged mechanically if the sample is heavy. (c) Cytotoxic materials will likely leach harmful substances that can damage the cells. The extent of damage can be evaluated by observing the cells' morphology, attachment, and absorption of "death stain" such as trypan blue.

TABLE 5.2

Scales for Cytotoxicity Evaluation for the Direct Contact Method

Scale	Observation
0	No significant change of cells in proximity to test sample
1	Small loss of cells and cellular deterioration beneath the sample
2	Cellular death and deterioration of all cells directly beneath the entire sample
3	Cellular death and deterioration moderately extended beyond the perimeter of test sample
4	Cellular death and deterioration greatly extended beyond the perimeter of the test sample

directly onto a monolayer of adherent cells. A mouse fibroblast cell line, L-929, is traditionally used. After a period of incubation, cells surrounding the test sample are observed for their morphology and viability, and the cytotoxicity of the test sample is evaluated qualitatively according to Table 5.2. For each testing, a positive control and a negative control must be included among the test samples. Suitable positive control materials are the ones that are known to be moderately cytotoxic such as organotin-stabilized polyvinyl chloride (PVC) and latex rubber, and suitable negative control materials are the ones that have no known *in vitro* cytotoxic effect, such as polyethylene (PE) or Teflon® (Dupont, Wilmington, Delaware).

To begin, we will seed L-929 cell cultures and allow them to grow to confluency. Each group will pick several different test materials for the cytotoxicity assays along with gum rubber (latex) as positive control and Teflon as negative control. These materials will be sterilized by autoclaving and then placed in direct contact with L-929 monolayers. After a period of incubation (24 hours or 48 hours), the cells will be stained with trypan blue and observed with microscopy. We will evaluate the cytotoxicity of the test materials according to Table 5.2.

Safety Notes

1. L-929 is a Biological Safety Level 1 (BSL-1) cell line, which means that it is not hazardous to *healthy* human individuals.
2. Trypan blue is a dye that is used for screening viable cells. ***Caution:*** It is carcinogenic and can be absorbed through skin. Gloves should be worn while working with trypan blue.

Procedures

Part 1 (Day 1). Preparing Cell Cultures

1. *Check-in*
   a. Samples and materials
      - L-929 cell culture in T-25 flask, near confluency (prepared by your group or your instructor)

   b. Reagents
      - [BSC] Complete MEM (with 10% FBS), warmed up to 37°C
      - [BSC] Trypsin Versene (Trypsin EDTA), warmed up to room temperature

  - [BSC]
  - Trypan blue

c. Special equipment and supplies
  - [BSC] 6-well tissue culture plate, sterile
  - [BSC] 15-ml conical tube, sterile

2. *Preparation*
   a. Turn on the blower of the biosafety cabinet at least 30 minutes ahead of time, and then wipe it down with 70% ethanol. Afterwards, aseptically move the items that are labeled with [BSC] in Step 1 into the biosafety cabinet.
   b. Observe the L-929 culture under the microscope and note the viability of the cells and the degree of confluency.
3. *Trypsinizing and suspending cells:* Trypsinize and suspend the L-929 cells according to standard procedures outlined below (see BLS II for details):
   a. Rinse cells with 5 ml of PBS.
   b. Incubate cells with 0.5 ml trypsin EDTA.
   c. Add 4.5 ml complete medium to suspend cells.
4. *Counting cells:* Determine the cell count of the suspension using standard procedures outlined below (see BLS II for details):
   a. Mix 20 µl of the cell suspension with 20 µl of trypan blue.
   b. Load the hemacytometer with the cell/dye mixture.
   c. Count the number of cells in 10 squares total.
   d. Calculate the cell count per ml:

$$\text{C.C.}\ \left(\text{ml}^{-1}\right) = N \times 10^3 \times 2$$

   where N is the total number of cells in 10 squares.

5. *Seeding a 6-well plate*
   a. *Note:* Use the worksheet in Table 5.3 for calculations in the following steps. We will seed cell cultures in a 6-well plate for the direct contact test. Before pipetting a cell suspension, always swirl it gently a few times to make sure that the cells are evenly suspended.
   b. Since we are seeding the 6-well plate several days ahead, we need to seed it with the appropriate number of cells so that it will reach confluency on the day of the direct contact test (Table 5.3, Section 1). Ask your instructor for the N_{cnfl} and the T_d for your L-929 cells, and calculate the target seeding number N_{seed}. (The confluency cell count and the doubling time for L-929 cells vary depending on the culture conditions, the reagents, and other factors. Use $N_{cnfl} = 3 \times 10^6$ and T_d = 22 hours as default if these parameters have not been specifically determined for your cultures.) *Note:* Cell cultures should not be seeded more than 5 days in advance. If the seeding density is too low, the cells will enter a lag phase in which they remain stationary for days before proliferating normally again.

TABLE 5.3

Worksheet for Cell Seeding

1. Determining the number of cells to seed:

Cell count at confluency N_{cnfl} = __________

Seeding ahead T = __________ days = __________ hours

Doubling time T_d = __________ hours

Number of cells to seed $N_{seed} = N_{cnfl} \times 2^{-\frac{T}{T_d}}$ = __________

2. Determining the volume of cell suspension needed for seeding:

The cell count of the cell suspension (CS) C.C. = __________ ml^{-1}

The volume of CS for N_{seed} cells $V_{cs} = N_{seed}/C.C.$ = __________ ml

Is V_{cs} <1.0 ml? Yes No

If yes, make diluted cell suspension (DCS): 1.0 ml of cell suspension + 9.0 ml medium

The volume of diluted cell suspension for N_{seed} cells $V_{DCS} = V_{cs} \times 10$ = __________ ml

3. Expanding V_{DCS} (or V_{cs}) to make cell seeding suspension (CSS):

The number of wells #W = __________ (6 for 6-well plate)

The volume to be added into each well V_{well} = __________ ml (2.0 ml for 6-well plate)

The volume of cell seeding suspension $V_{css} = \#W \times V_{well} + 0.5$ ml = __________ ml

To make CSS: Mix V_{DCS} of DCS with medium to a total volume of V_T

(or if V_{CS} >1.0 ml, then mix V_{cs} of CS with medium to a total volume of V_T)

To seed the cultures: Distribute a volume V_{well} of CSS to each well.

c. Based on the cell count of the cell suspension from Step 3, calculate the volume, V_{cs}, needed for the target seeding number (Table 5.3, Section 2). If V_{cs} is <1.0 ml, it is difficult to pipette accurately, and a diluted cell suspension (DCS) should be made by mixing 1.0 ml of the cell suspension from Step 3 and 9.0 ml of complete α-MEM in a 15-ml tube. Label it as "DCS." The volume for the DCS will then be V_{Dcs}.

d. To evenly distribute the cells from DCS into the wells of 6-well plate, we need to expand the volume of the DCS for convenient and accurate pipetting (Table 5.3, Section 3). For 6-well plate, exactly 12.0 ml is needed for 2.0 ml in each of the six wells, but we add an extra 0.5 ml to accommodate pipetting errors. (Because in this case, it is more important for all six wells to have the same number of cells; we want to avoid not having enough cell suspension for the sixth well.) Add V_{DCS} of DCS to a 15-ml conical tube, and then add α-MEM to a total volume of 12.5 ml; label it as "CSS" for cell seeding suspension.

e. Finally, dispense 2.0 ml of the CSS to each well of the 6-well plate. Label the plate with the name of the cell line (L-929), your group's name, the date, and the seeding cell count (N_{seed}). Move the plate to a 37°C CO_2 incubator.

6. *Finishing up:* Store the reagents properly. Make sure that all disposable items are properly discarded. Wipe down the biosafety cabinet.

Part 2 (Day 1). Preparing Test Materials

1. *Check-in*

   a. Samples and materials

- Assorted materials for biocompatibility testing; possible testing materials include but not limited to:
  - PTFE (Teflon), negative control
  - Gum rubber (latex), positive control
  - Stainless steel
  - Titanium
  - Hydroxyapatite
  - Silicone rubber
  - Amalgum
  - Vinyl
  - Chromium
  - Neoprene

b. Special equipment and supplies
  - Autoclave pouches

2. *Sterilizing the test materials*

   a. Pick the negative and positive controls and four other materials for the direct contact tests. Seal each sample in *its own* sterilization pouch, and mark it with the name of the material, your group's name, and today's date.

   b. Autoclave the samples at 120°C and 15 psi for >30 minutes. *Note:* Make sure that the materials you have picked are autoclavable; if not, other methods must be used to sterilize them.

Part 3 (Day 2). Placing Test Materials in Direct Contact with Cell Cultures

1. *Check-in*

   a. Samples and materials
      - L-929 cell cultures in 6-well plate, near confluent (from Part 1)
      - [BSC] Sterilized test materials (from Part 2)

   b. Reagents
      - [BSC] Complete MEM, warmed up to 37°C

   c. Special equipment and supplies
      - [BSC] Forceps or tweezers, autoclaved

2. *Preparation*

   a. Vent the BSC for 30 minutes and wipe it down with 70% ethanol. Aseptically move the items that are labeled with [BSC] in Step 1 into the BSC.

   b. Observe the L-929 cell cultures under the microscope. Make a note of the viability and the confluency of cells.

   c. Move the 6-well plate aseptically into the biosafety cabinet. Assign one well to each test material and label the assignment on the lid.

3. *Setting up the direct contact*

   a. Aspirate the old medium from the wells, and add 0.5 ml complete MEM back to each well.

b. Open the sterilization pouches containing the test materials. Place the test materials gently in the middle of the corresponding wells using a pair of forceps. Try not to scratch the cells. If a material tends to float, weigh it down with a piece of Teflon, the positive control material.
c. Add another 0.5 ml complete MEM to each well; gently deliver the medium against the walls of the wells to avoid disturbing the samples.
d. Label the plate with your group's name and the date. *Carefully* move it to a 37°C CO_2 incubator.
e. Incubate the cell cultures for 24 to 48 hours (according to the class schedule).

4. *Finishing up:* Discard the empty pouches, store the reagents properly, and wipe down the biosafety cabinet.

Part 4 (Day 3). Evaluation of Cytotoxicity

1. *Check-in*
   a. Samples and materials
      - [BSC] L-929 cell cultures with test materials in direct contact
   b. Reagents
      - [BSC] Complete MEM, warmed up to 37°C
      - Trypan blue
   c. Special equipment and supplies
      - [BSC] Forceps or tweezers, autoclaved
      - Container for trypan blue liquid waste
2. *Preparation:* Vent the BSC for >30 minutes and wipe it down with 70% ethanol. Aseptically move the items that are labeled with [BSC] in Step 1 into the BSC. Be careful not to disturb the test materials when moving the 6-well plate.
3. *Staining the cells with trypan blue*
   a. Remove the test samples from the cell cultures with a pair of forceps. Avoid scratching the cells.
   b. Aspirate the old medium, and then add 0.5 ml of fresh media to each well.
   c. Move the plate out of the biosafety cabinet. Add 0.5 ml of trypan blue to each well and swirl the plate to mix the dye. *Note:* From this point on, it is no longer necessary to keep the cell cultures in the plate aseptic since they will not be cultured further.
4. *Evaluation*
   a. *Note:* We will observe the viability of the cells, especially those around the test materials, to determine the cytotoxicity of the test materials. Healthy cells should adhere to the surface. Detachment can be detected by gently agitating the plate—detached cells will be floating around. Morphologically, healthy cells tend to be spindle shaped and have protrusions from the cell body for adhesion (see Figure 4.6). You should also calibrate your observations by comparing the positive and negative controls.
   b. Secure the plate on the mechanical stage of the microscope. Use phase contrast microscopy to observe the shapes of the cells and brightfield microscopy

to observe the stained cells, and try different color filters to enhance the brightfield views. Use various magnifications for overviews and detailed views of the cells around the test material. Take both phase contrast and brightfield images for each test material. (See Data Processing and Figure 5.3 for image requirements.)

5. *Finishing up*
   a. Save or transfer the image files to the computer. Last users should turn off the microscope.
   b. Pipette the trypan blue solution in the 6-well plate to the trypan blue waste container; spray 70% ethanol to rinse the wells, and then pipette off the rinse to the waste container as well. After rinsing, discard the 6-well plate as bio-hazard waste.
   c. Store the reagents properly and wipe down the biosafety cabinet.
   d. Collect the test materials in a beaker and soak them with 70% ethanol. The polymer materials such as gum rubber and neoprene can be discarded, while the (expensive) metal materials such as titanium and stainless steel can be re-used.

Data Processing

1. *Data from your own group*
   a. Evaluate the cytotoxicity of the test materials according to the scales in Table 5.2. Use the positive and negative controls to decide the validity of the results. The experiment is deemed unsuccessful if the control materials do not yield expected results, and results for the test materials will become invalid. The cytotoxicity scale should be 0 or 1 for the negative control, and 2 or 3 for the positive control.
   b. Use images of the cells to support your evaluation. You should use phase contrast images to show the morphology of the cells, and brightfield images to show trypan blue staining (or the lack of it). Each image must include a scale bar; important features such as a zone of dead cells should be marked in the image and described in the figure legend (Figure 5.3).
2. *Data from class:* The positive and negative controls help us to check for system errors in the experiment, and on the other hand, parallel experiments are needed to calibrate the range of random errors. In order for your cytotoxicity evaluations to be statistically significant, you must include results from the other groups in class. (If there is no other group, then you should perform your experiment in (at least) duplicate.) List the cytotoxicity scales obtained by other groups along with your own, and qualitatively decide whether your evaluations are reliable or not reliable:
   - Reliable: when all results for a particular material are different by no more than one scale unit.
   - Not reliable: when they are different by two or more scale units.

For example, for a particular test material, if group A (your group) decides the cytotoxicity scale is 1, group B gives it a 2, and group C gives it a 3, then your conclusion, a cytotoxicity scale of 1, should be deemed not reliable.

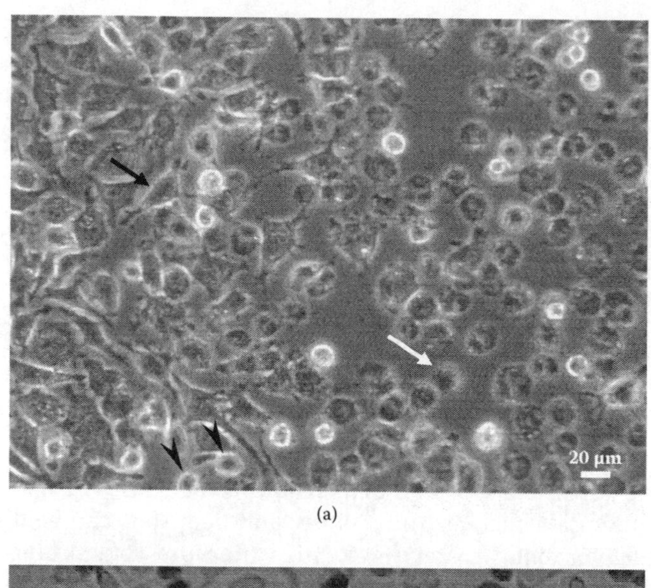

(a)

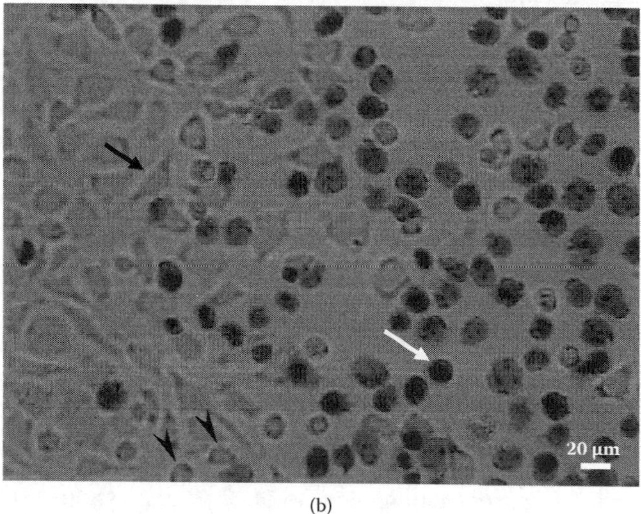

(b)

FIGURE 5.3
Examples of images from a direct contact test. Each image should have a scale bar length that represents a, and important features should be pointed out. The scale bar for these two images is 20 μm. The images are taken around the original location of a positive control material. Viable cells maintain their spindle shape, and they are not stained with trypan blue (long black arrows), whereas nonviable cells are round and stained (long white arrows). On the other hand, dividing cells or just-divided cells tend to be round in shape (black arrow heads), but not stained with trypan blue since they are viable. (a) Phase contrast image. (b) Brightfield image of the same view as A.

Session 2. Cytotoxicity Evaluation Using Liquid Extracts of Materials

In addition to the direct contact test, another widely used *in vitro* cytotoxicity testing method is the extraction test (or the elution test). In this test (Figure 5.4), the test material is extracted with a liquid, which can be cell culture medium, saline solution, vegetable oil, or another solvents that are relevant to the application of the test material. According to

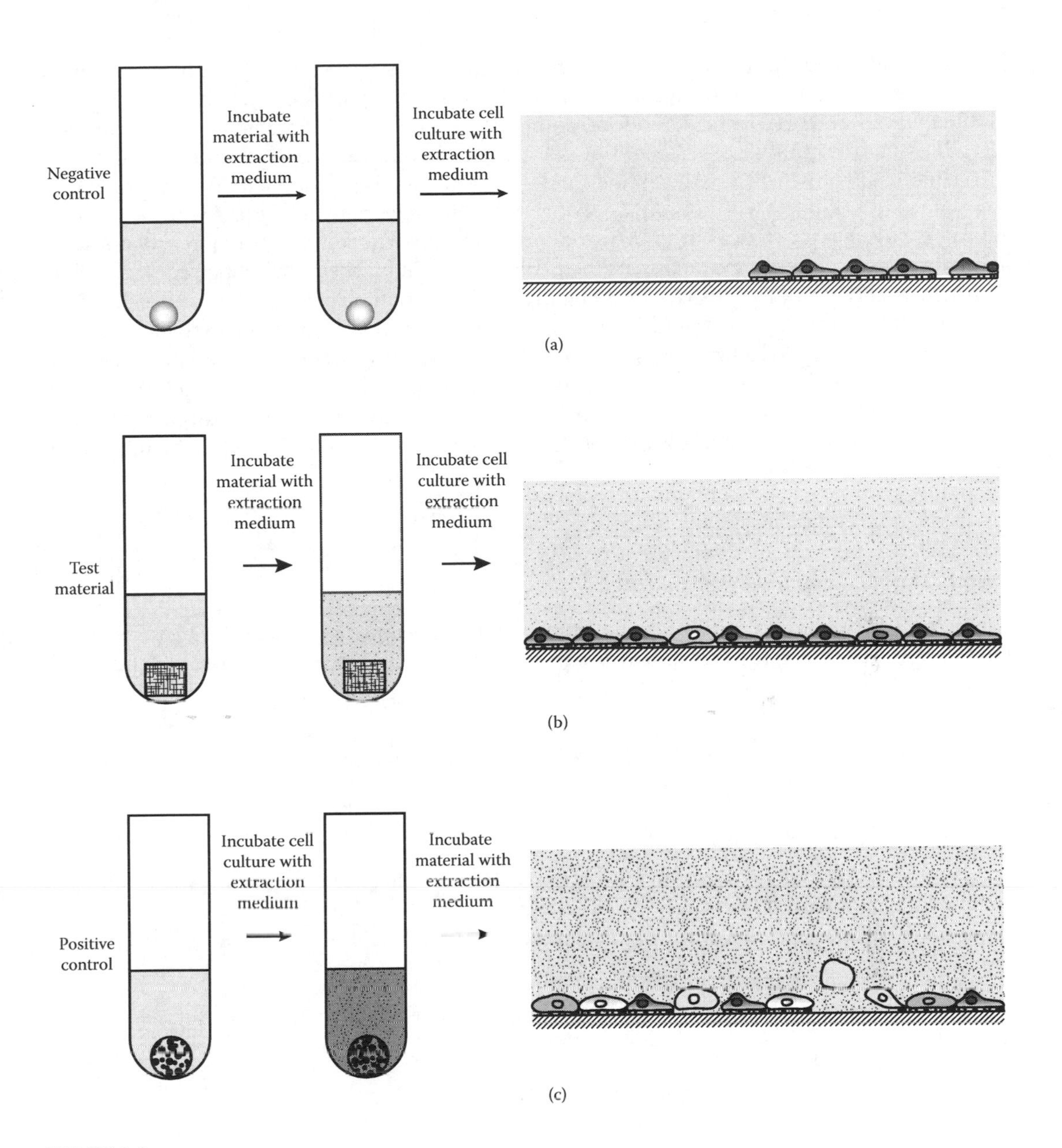

FIGURE 5.4
Cytotoxicity evaluation of material using the extraction test. The test material (b) is extracted with an appropriate solution or solvent along with a negative (a) and positive control (c). After incubating cell cultures with the extracts for a period of time, the cytotoxicity of the test material can be evaluated qualitatively as well as quantitatively by comparing the percentage of viable cells in the test culture to those in the control cultures.

the ISO 10993-5 guidelines (ISO 1999), "extraction conditions should attempt to simulate or exaggerate the conditions of clinical use so as to determine the potential toxicological hazard, without causing significant changes in the test material such as fusion, melting or alteration of the chemical structure." The extract is then added to cell cultures, and its effect on the viability of the cells will indicate the cytotoxicity of the material. For comparison, positive and negative control materials are extracted and tested under the same conditions. If the *in vitro* cell cultures have the same number of cells to begin with, then after incubating with the extract, the viability of the cells tested with the positive control and the test materials can be compared to that with the negative control by counting the viable cells in all three tests. Therefore, compared to the direct contact test, the extraction test allows a quantitative assessment of cytotoxicity in addition to a qualitative one.

In this session, we will use the extraction test to evaluate the cytotoxicity of a tubing material. The extraction test would be an appropriate test if we were to determine whether certain tubing is suitable for clinical fluid transfer, for example, because the extraction process actually mimics the constant extraction of the tubing material by the fluid that circulates inside. The test material and the positive and negative controls will be extracted in complete MEM at a ratio of 2.0 cm^2 surface area to 1.0 ml medium for a period of time. The extracts will then be added to duplicate cell cultures in a 6-well plate. After a period of incubation, we will count the number of viable cells in each culture. Since the plate was seeded such that all six wells have the same number of cells before incubation, by comparing the numbers of viable cells after incubation, the cytotoxicity of the test material can be evaluated according to the cell viability percentage.

Procedures

Part 1 (Day 1). Preparing Cell Cultures

Prepare L-929 cell cultures in a 6-well plate using the same procedures as in Session 1, Part 1. (Alternatively, the cell cultures can be prepared by the instructor for better seeding consistency.)

Part 2 (Day 1). Sample Preparation for the Extraction Tests

1. *Check-in*
   a. Sample and materials
      - Test materials for the extraction tests; possible testing materials include but not limited to:
        - Gum rubber tubing (positive control)
        - Teflon tubing (negative control)
        - Silicone tubing, outer diameter (OD) ~1.2 cm
        - Tygon tubing, OD ~1.2 cm
        - Neoprene tubing, OD ~1.2 cm
   b. Special equipment and supplies
      - Sterilization pouches
      - Micrometer
2. *Preparation*
   a. *Note:* We will test three materials: a positive control, a negative control, and a material to be tested for cytotoxicity.
   b. Cut a ~1.5 cm segment from each tubing. For each segment, make triplicate measurements of the length, inner diameter, and outer diameter.
   c. Wash the samples with de-ionized water, and then seal them in separate sterilization pouches. Label each pouch with the name of the material, your group's name, and the date. Autoclave the samples at 120°C and 15 psi for >30 minutes.

d. *Homework:* Calculate the surface area of all three samples, and then calculate the volume of media needed for extraction of each of the sample using a ratio of 2 cm^2/ml. (You will need these numbers for Part 3.)

Part 3 (Day 2). Extracting the Test Materials

1. *Check-in*
   a. Samples and materials
      - [BSC] Sterilized positive control, negative control, and your choice of one test material.

   b. Reagents
      - [BSC] Complete MEM, warmed up to 37°C

   c. Special equipment and supplies
      - [BSC] Three 15-ml tubes, sterile
      - [BSC] Tweezers, autoclaved
2. *Preparation:* Vent the BSC and wipe it down with 70% ethanol. Aseptically move the items labeled with [BSC] in Step 1 into the BSC.
3. *Extraction*
   a. Use a pair of tweezers to transfer each test sample into a 15-ml tube, and then add complete MEM into the 15-ml tubes according to the volumes that you have calculated in Part 2, Step 2d. Label each 15-ml tube with the name of the test material, your group's name, and the date.

   b. Incubate the materials with extraction medium in a 37°C CO_2 incubator for ~24 hours. Store the tubes at 4°C afterwards. (The test materials can be left inside the tubes.)
4. *Finishing up:* Discard the empty sterilization pouches, store the medium, and wipe down the biosafety cabinet.

Part 4 (Day 3). Incubating Cell Cultures with Extracts

1. *Check-in*
   a. Samples and materials
      - [BSC] Extracts of the control materials and the test material, warmed up to 37°C
      - [BSC] L-929 cultures in a 6-well plate (from Part 1), near confluent
2. *Preparation*
   a. Observe the cell cultures under the microscope and note the confluency and viability of the cells.

   b. Vent the BSC and wipe it down with 70% ethanol. Aseptically move the test materials and the cell cultures inside the BSC.

   c. Assign two wells each in the 6-well plate for the positive control, the negative control, and the test material, and label the assignment on the lid of the plate. *Note:* Each material will be tested with duplicate cultures.

3. *Incubating cell cultures with the extracts*
   a. Aspirate the old medium from the cell cultures in the 6-well plate, and then add 2.0 ml of the extract for each material to the corresponding wells. *Note:* If not enough extract is available, use a smaller volume than 2.0 ml, but the volume should be the same in each well; for each material, make sure that you have enough extract for duplicate wells.
   b. Move the cell cultures to a 37°C CO_2 incubator and incubate for 24 to 48 hours (depending on the class schedule).
4. *Finishing up:* Discard the 15-ml tubes with extracted material samples into biohazard waste. Wipe down the biosafety cabinet.

Part 5 (Day 4). Evaluation of Cytotoxicity

1. *Check-in*
   a. Samples and materials
      - [BSC] L-929 cultures in 6-well plate incubated with extracts of the test material and the control materials
   b. Reagents
      - [BSC] Complete MEM, warmed up to 37°C
      - [BSC] Trypsin Versene (trypsin EDTA)
      - [BSC] Phosphate-buffered saline (PBS)
      - Trypan blue
   c. Special equipment and supplies
      - Hemacytometers
      - Compressed air (from cylinder or duster)
      - Ice (or 4°C refrigerator)
      - [BSC] One 1.5-ml microcentrifuge tubes, autoclaved
      - [BSC] One 15-ml conical tube, sterile
      - [BSC] A 200-μl micropipette
2. *Preparation*
   a. Observe the cell cultures under the microscope and take notes of the morphology and attachment of the cells in each well. Note that viable cells should have spindle-like shapes and attach to surface, and nonviable cells are usually round and detached.
   b. Vent the BSC and wipe it down with 70% ethanol. Aseptically move the items labeled with [BSC] in Step 1 into the BSC.
3. *Trypsinizing and suspending cells*
   a. In the BSC, pipette ~1.3 ml trypsin into a sterile 1.5-ml microcentrifuge tube to be used in the next step.
   b. Aspirate the medium from all 6 wells and rinse each well with 1 ml PBS. Aspirate the PBS. Then using the 200-μl micropipette, add 200 μl trypsin EDTA to each well. Rock the plate to coat the cell surfaces evenly with trypsin. *Note:* Here we use a micropipette for volume accuracy, which is critical for accurate cell counting.

c. Incubate the plate in a 37°C CO_2 incubator for ~2–3 minutes. Meanwhile, pipette ~6.5 ml of complete MEM to a 15-ml conical tube, and then move the tube out of the BSC. *Note:* Procedures from this point on will be carried out outside the BSC.

d. Observe the trypsinization under the microscope. Tap the 6-well plate to make sure that all cells are dislodged. *Note:* Incomplete trypsinization will result in big errors in the cell counts, so make sure that the cells are completely detached before the next step.

e. Using a 1000-µl micropipetter, add 1000 µl complete MEM from the 15-ml tube into each well. Pipette up and down in each well to suspend the cells, and then transfer ~0.5 ml of the cell suspension from each well to a 1.5-ml microcentrifuge tube. Label the microcentrifuge tubes and place them on ice. *Note:* Use a new pipette tip for cell suspending in each well.

4. *Cell counting*

   a. *Important:* Always mix the cell suspension before pipetting since the cells tend to sediment.

   b. In a 0.5-ml or 1.5-ml microcentrifuge tube, add 20 µl trypan blue, and then mix in 20 µl of cell suspension.

   c. Load the cell/dye mixture into a hemacytometer and count the cells using standard procedures.

   d. Repeat the counting procedure for the rest of the samples. Clean the hemacytometer each time before loading a new sample. If available, your group can use two or more hemacytometers to do the cell counting more efficiently.

5. *Finishing up:* Clean the hemacytometers. Discard the 6-well plate and other disposable items into biohazard waste. Store the reagents and wipe down the biosafety cabinet.

Data Processing

1. *Data from your own group:* Calculate the relative cell count of the test material and the positive control as a percentage of the negative control (Teflon). You should have two data points for each material since you performed the test in duplicates.
2. *Data from class:* Pool the relative cell counts obtained by other groups in the class. For each test material and the positive control, calculate the average relative cell count and the standard error, respectively.
3. Evaluate the cytotoxicity scale of the test material using the scales in Table 5.4. For a reliable evaluation, the positive control should have a cytotoxicity scale of at least 2, and the standard error for the test material should be ≤20%—the average range of the relative cell count for each increment of the cytotoxicity scale in Table 5.4.

TABLE 5.4

Cytotoxicity Assessment Using Extraction Test

Scale	Cell Count Comparison
0	Test plate growth estimated as 90% or more of control growth
1	Test plate growth estimated as 70–90% of control growth
2	Test plate growth estimated as 40–70% of control growth
3	Test plate growth estimated as 10–40% of control growth
4	Test plate growth observed to be minimal (less than 10% of control growth)

Session 3. Biofunctionality Evaluation through Cell Adhesion/Proliferation Tests

As mentioned previously, for many biomedical applications, biocompatibility requirements for the materials include both biosafety and biofunctionality. Criteria for biofunctionality are specific to the application. For example, the biofunctionality of a bone-replacement material is determined by how well it supports cell attachment and stimulates bone ingrowth; on the other hand, the biofunctionality of a good skin-replacement material is determined by how well it promotes skin cell growth, and vascularization. Biofunctionality can be evaluated by criteria such as cell adhesion, cell spreading, cell proliferation, biosynthesis, production of specific molecules by the cells, etc. Some of these studies are applicable for most materials, and others are tailored for specific applications. In this session, we will study the interaction between cells and the material surface by studying cell proliferation on the surfaces of potential bone-replacement materials. Specifically, we will study the proliferation of preosteoblast cells on titanium or stainless steel (Figure 5.5). In this experiment, cell proliferation is also an indication for cell adhesion: better adhesion leads to better proliferation. For comparison, we will also grow cells on Teflon, to which cells adhere poorly due to its hydrophobic surface.

The preosteoblast cells that we will use in our experiment are from a mouse predifferentiation osteoblast cell line, MC3T3-E1. This cell line simulates the physiological behaviors of osteoblasts: It secrets collagen, and when induced with ascorbic acid, it differentiates and forms a mineralized extracelluar matrix that has a similar composition as bone. Therefore, the proliferation behaviors of MC3T3-E1 cells will provide very good evaluation of the biofunctionality of the test materials as prospective bone-replacement materials.

To evaluate the proliferation of MC3T3-E1 cells quantitatively, we will use a colorimetric assay based on MTT [3-(4,5-dimethylthiazol-2-yl)-2,5-diphenyl tetrazolium bromide]. When MTT is added to metabolically active cells, it is cleaved to form a dark purple formazan product (Figure 5.6). MTT formazan forms water-insoluble crystals that can be solubilized by dimethylsulfoxide (DMSO) and alcohols. In spectrophotometry, MTT formazan solution has a maximum absorption at 570 nm. MTT assay is especially sensitive for measuring cell proliferation: proliferating cells are more metabolically active compared to nonproliferating cell, thus proliferating cells convert MTT more actively; dead cells or metabolically inactive cells do not react with MTT at all. Moreover, the amount of MTT formazan produced is directly proportional to the number of proliferating cells present in a culture; as a result, calibration curves such as the one in Figure 5.7 can be established and used for cell counting. Because the metabolic activity is different for different cell types (Mosmann 1983), the calibration curve must be cell-type specific in order to accurately convert MTT formazan absorbance to cell count. To establish a calibration curve, MTT is typically added to a series of cell cultures with known numbers of cells, and the MTT formazan absorbance measurements are then plotted versus the numbers of cells. The data points are fitted with a linear equation (Figure 5.7). Using this equation, the number of cells for a given culture can be calculated after measuring the absorbance of MTT formazan produced by the cells. In practice, typical steps for an MTT assay with calibration are outlined in Figure 5.8.

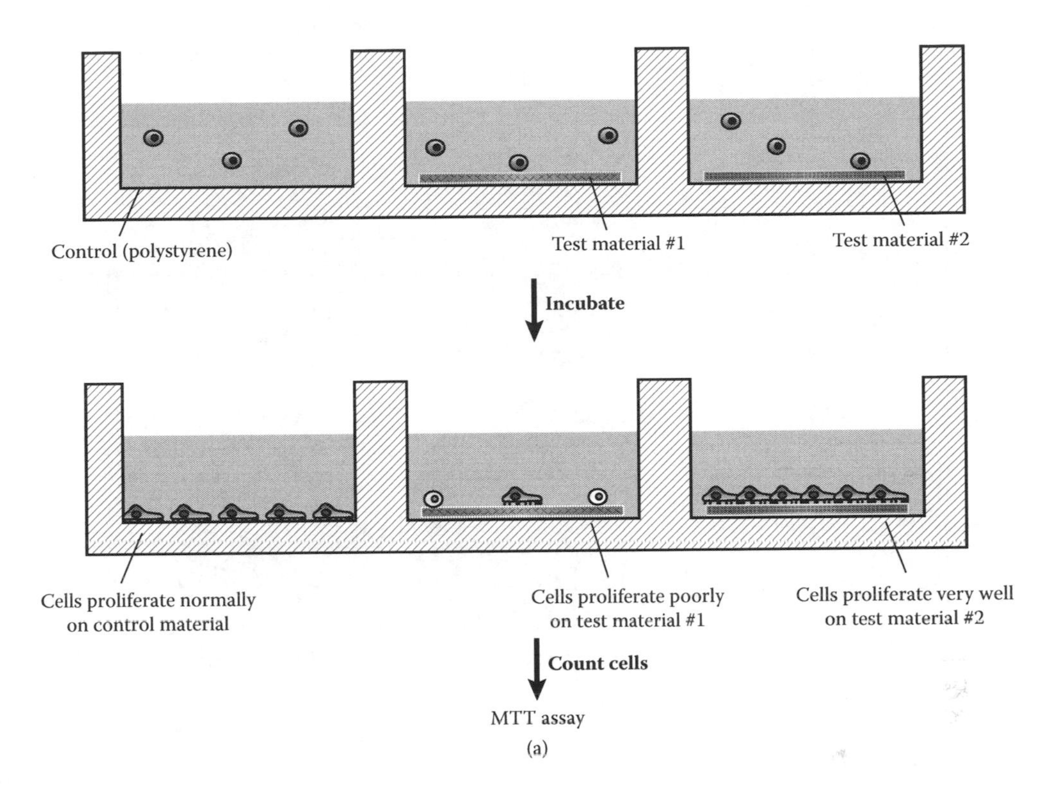

FIGURE 5.5
Biofunctionality evaluation of material by studying cell adhesion and the subsequent cell proliferation. (a) Cells are seeded on the surfaces of test materials and a positive control material that support cell growth. Cell proliferation during the subsequent incubation period is partly determined by adhesion of the cells to the surface of material: cells adhere poorly to test material #1 and do not proliferate, but they adhere strongly to test material #2 and proliferate vigorously as a result. Cell proliferation is then measured quantitatively with the MTT assay, an assay based on the chemical conversion of MTT to MTT formazan by proliferating cells. (b) Example: Arrangement of samples in a 6-well plate for cell proliferation study. Test materials are cut into discs and placed in the wells. The plasma-treated polystyrene well surface is used as positive control to which the cells will adhere strongly. The test materials and the control are measured in duplicate. Note that the surface area difference between the test material discs and the wells should be taken into account when comparing cell counts.

Procedures

Part 1 (Day 1). Preparing Test Materials

1. *Check-in*
   a. Samples and materials
      - Two titanium (or stainless steel) discs, ~1″ in diameter
      - Two Teflon discs, ~1″ in diameter
   b. Special equipment and supplies
      - Micrometer
      - Sterilization pouches
      - Tweezers

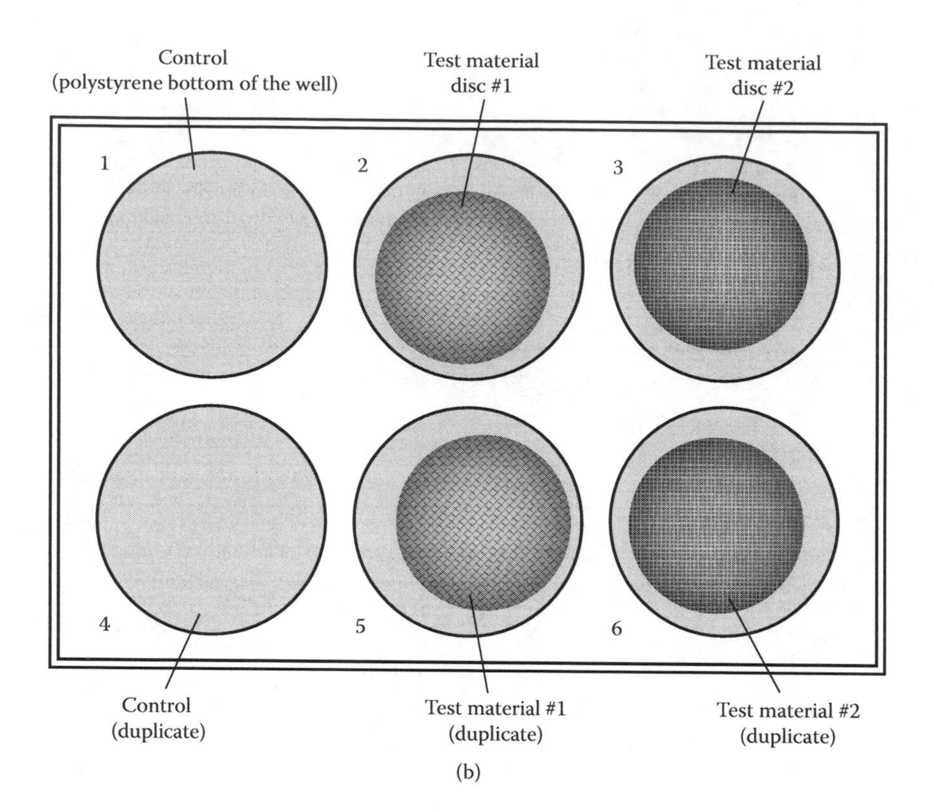

FIGURE 5.5 (continued)

2. *Preparation*
   a. *Note:* Titanium (or stainless steel) discs with ~1″ diameter should be cut from sheets 0.1–0.2 mm thick. Teflon discs, also ~1″ diameter, should be cut from sheets 0.3–0.5 mm thick.
   b. Use a micrometer to measure the exact diameters of the discs. Also measure the diameter of the wells of the 6-well plate that you will use in Part 2 (or obtain it from the manufacturer). All measurements should have an accuracy of 0.1 mm or better. Tabulate the measurements in your lab notebook.
   c. Wash the discs with dish detergent, and then rinse them with water and de-ionized water. Dry the discs with Kimwipe tissue paper.
   d. Seal the discs in a sterilization pouch and label the pouch. Autoclave the discs at 120°C and 15 psi for >30 minutes.

Part 2 (Day 2). Seeding MC3T3-E1 Cells on the Surfaces of Test Materials

1. *Check-in*
   a. Samples and materials
      - [BSC] Stainless steel or titanium discs from Part 1, autoclaved
      - [BSC] Teflon discs from Part 1, autoclaved
      - [BSC] MC3T3-E1 cultures in T-25 flasks

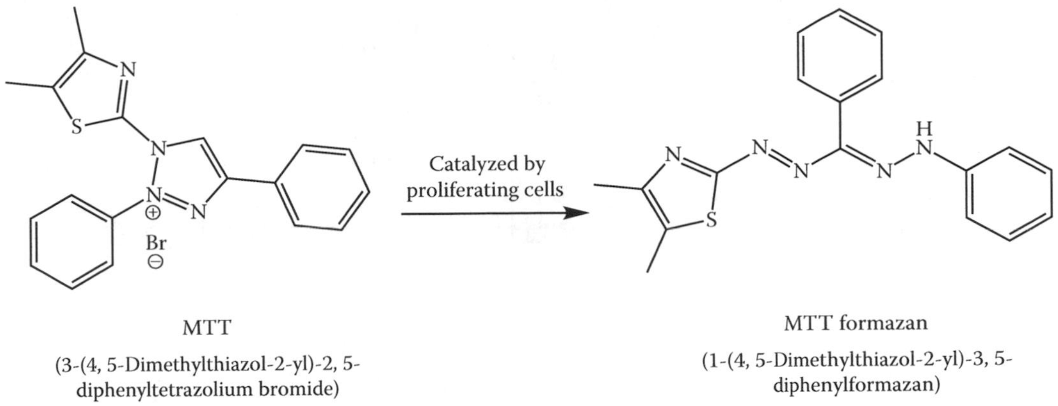

N
S
N
N
N
Br
Catalyzed by
proliferating cells
N
S
N
N
N
H
N
MTT
(3-(4, 5-Dimethylthiazol-2-yl)-2, 5-diphenyltetrazolium bromide)
MTT formazan
(1-(4, 5-Dimethylthiazol-2-yl)-3, 5-diphenylformazan)

FIGURE 5.6
The chemical reaction that converts MTT to MTT formazan, a compound with a dark purple color that is insoluble in water.

y = 6.8518 × 10^-6 x + 0.14227
Absorbance (570 nm)
0
0.5
1
1.5
2
0
50000
100000
150000
200000
250000
Number of Cells (MC3T3-E1)

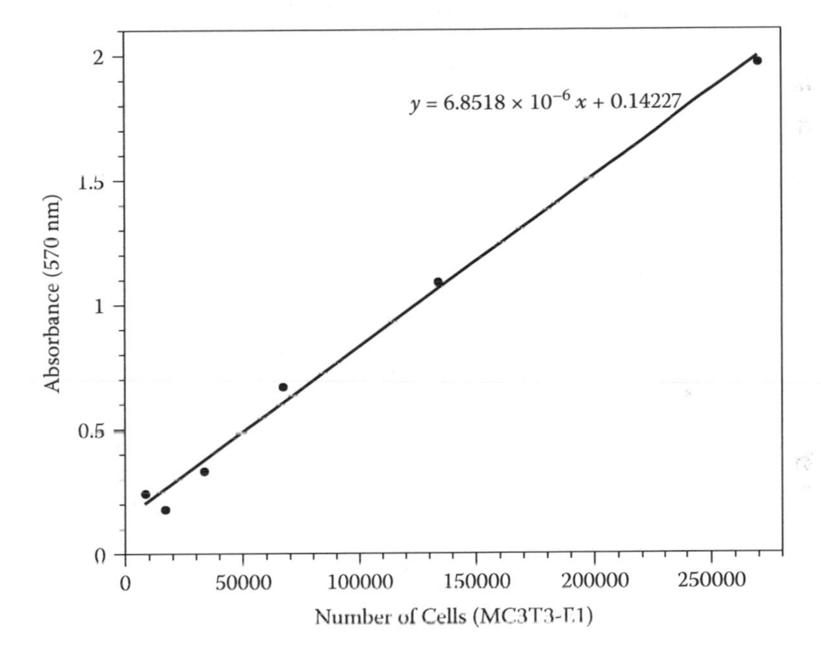

FIGURE 5.7
MTT assay calibration curve for MC3T3-E1 cells. Note that the curve can be fitted very well with a linear equation.

b. Reagents
   - [BSC] Complete α-modified minimum essential medium (α-MEM), prepared with ribonucleosides, deoxyribonucleosides, 2 *mM* L-glutamine, 1 *mM* sodium pyruvate, 1% penicillin/streptomycin and 10% FBS, warmed up to 37°C
   - [BSC] Trypsin EDTA, warmed up to room temperature
   - [BSC] PBS
   - Trypan blue

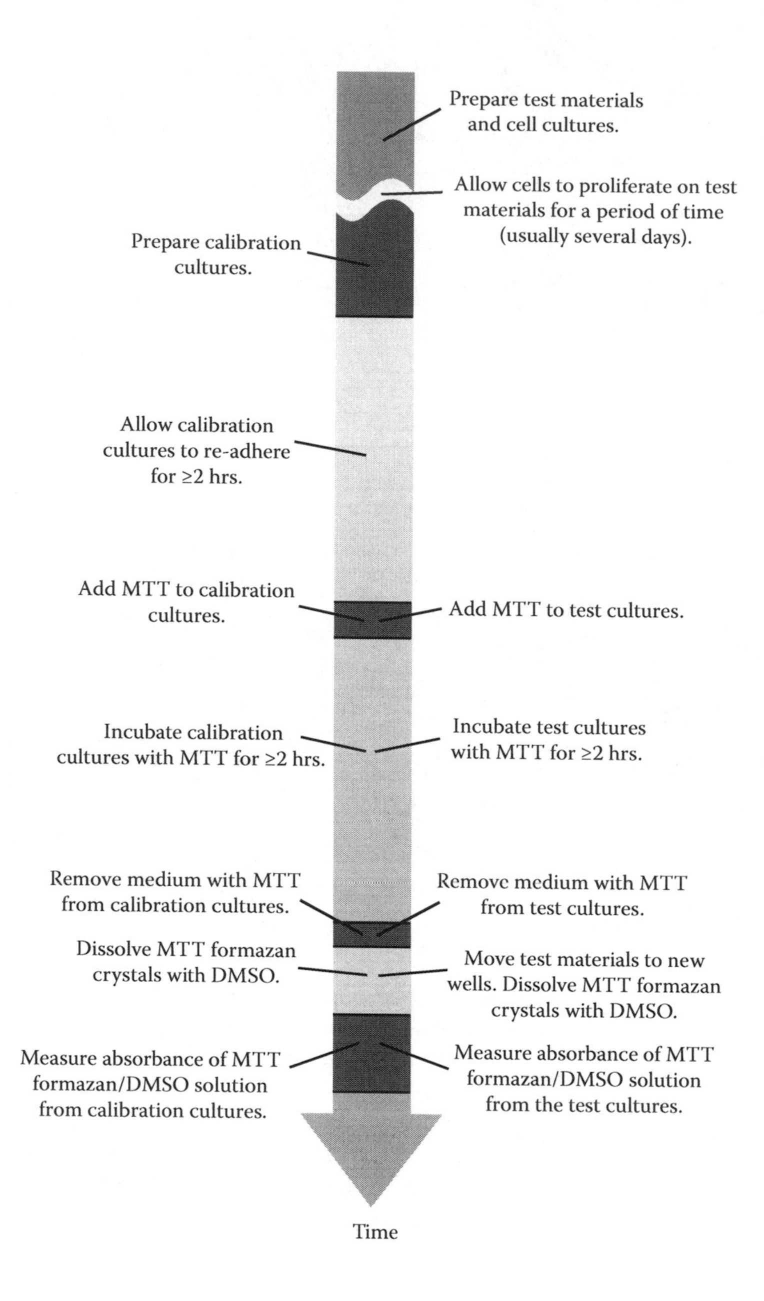

FIGURE 5.8
Steps in measuring cell proliferation on material surface using MTT assay with calibration. Approximate blocks of time needed for the steps are arranged by order on a time axis. Steps for the calibration are labeled on the left of the time axis, and those for the testing are labeled on the right. Some of steps for the calibration and the testing can be performed at the same time.

c. Special equipment and supplies
   - Hemacytometer
   - [BSC] One 6-well plate, sterile
   - [BSC] Two 15-ml conical centrifuge tubes, sterile
   - [BSC] Tweezers, autoclaved

2. *Preparation*
    a. Observe the MC3T3-E1 culture under the microscope and note the viability of the cells and the degree of confluency. Especially pay attention to the morphology of MC3T3-E1 cells and see how it is different from that of L-929 cells.
    b. Vent the BSC and wipe it down with 70% ethanol. Aseptically move the items that are labeled with [BSC] in Step 1 into the BSC.
    c. Using a pair of forceps, place the four discs in the 6-well plate, one in each well. The two empty wells will be used as controls. Label each well on the lid with the name of the test material inside.
3. *Seeding MC3T3-E1 cells:* Seed MC3T3-E1 cells into the wells of the 6-well plate from Step 2 according to Session 1, Part 1 and Table 5.3, except that for MC3T3-E1 cells, use complete α-MEM as medium, and use $N_{cnfl} = 1.8 \times 10^6$ for the 6-well plate and $T_d = 20$ hours as default if these parameters have not been specifically determined for your cultures.
4. *Finishing up:* Clean the hemacytometers. Discard the disposable items properly, store the reagents, and wipe down the biosafety cabinet.

Part 3 (Day 3). Preparing MC3T3-E1 Cultures for MTT Calibration

(Optionally, this part can be performed by the instructor ahead of Part 4 on the same day. Students: You should understand the following procedures even if you do not need to perform the experiment yourselves.)

1. *Check-in*
    a. Samples and materials
        - [BSC] MC3T3-E1 culture in T-25 flask, near confluent
    b. Reagents
        - [BSC] Complete α-MEM, warmed up to 37°C
        - [BSC] Trypsin EDTA, warmed up to room temperature
        - [BSC] PBS
        - Trypan blue
    c. Special equipment and supplies
        - [BSC] One 6-well plate, sterile
2. *Preparation:* Vent and wipe down the BSC. Aseptically move the items that are labeled with [BSC] in Step 1 into the BSC.
3. *Suspending MC3T3-E1 cells:* Use the standard procedures outlined below:
    a. Rinse the cells with 5 ml of PBS.
    b. Add 0.5 ml trypsin EDTA. Incubate for 2 minutes.
    c. Add 4.0 ml complete α-MEM to suspend the cells.
4. *Counting the cells:* Use standard procedures, which are outlined below:
    a. Mix 20 µl of MC3T3-E1 cell suspension with 20 µl of trypan blue.
    b. Load the cell-dye mixture onto the hemacytometer, and count cells in 10 squares.

c. Calculate the cell count:

$$\text{C.C.}\ \left(\text{ml}^{-1}\right) = N \times 10^3 \times 2$$

where N is the total number of cells in 10 squares.

5. *Seeding cells with serial dilution*
   a. *Note:* We will use a serial dilution method (Figure 5.9) to establish a series of cell cultures with decreasing numbers of cells.
   b. Label the wells in the 6-well plate sequentially on the lid (or use the number molded on the upper-left corner of each well).
   c. Add 2.0 ml of complete α-MEM each in wells #2–6.
   d. First, add 4.0 ml of the MC3T3-E1 cell suspension (from Step 3) to well #1. Next pipette 2.0 ml from well #1 to well #2 and mix thoroughly; next pipette 2.0 ml from well #2 to well #3, and so on until the last 2.0 ml is removed from well #6. When finished, each well should have 2.0 ml of cell suspension, but the number of cells is half of that in the previous well. *Note:* For each transfer, make sure that the cell suspension is well mixed with the medium before transferring to the next well.
   e. Move the 6-well plate to a 37°C CO_2 incubator. Allow ≥2 hours for the cells to re-adhere to the surface.
   f. Calculate and tabulate the cell count in each well.

Part 4 (Day 3). MTT Assay

Caution: MTT is considered toxic. Handle with gloves.

1. *Check-in*
   a. Samples and materials
      - [BSC] MC3T3-E1 culture with test materials in 6-well plate from Part 2
      - [BSC] MC3T3-E1 calibration cultures prepared in Part 3
   b. Reagents
      - [BSC] Complete α-MEM, warmed up to 37°C
      - [BSC] PBS
      - [BSC] MTT dissolved in PBS at a concentration of 5 mg/ml, sterilized with 0.22 μm filtration.
      - DMSO
   c. Special equipment and supplies
      - [BSC] 200-μl pipette tips with filters, sterile (or autoclaved)
      - [BSC] 100-μl or 200-μl micropipette
      - One 6-well plate
      - Fine-tipped jeweler's forceps
      - Shaker

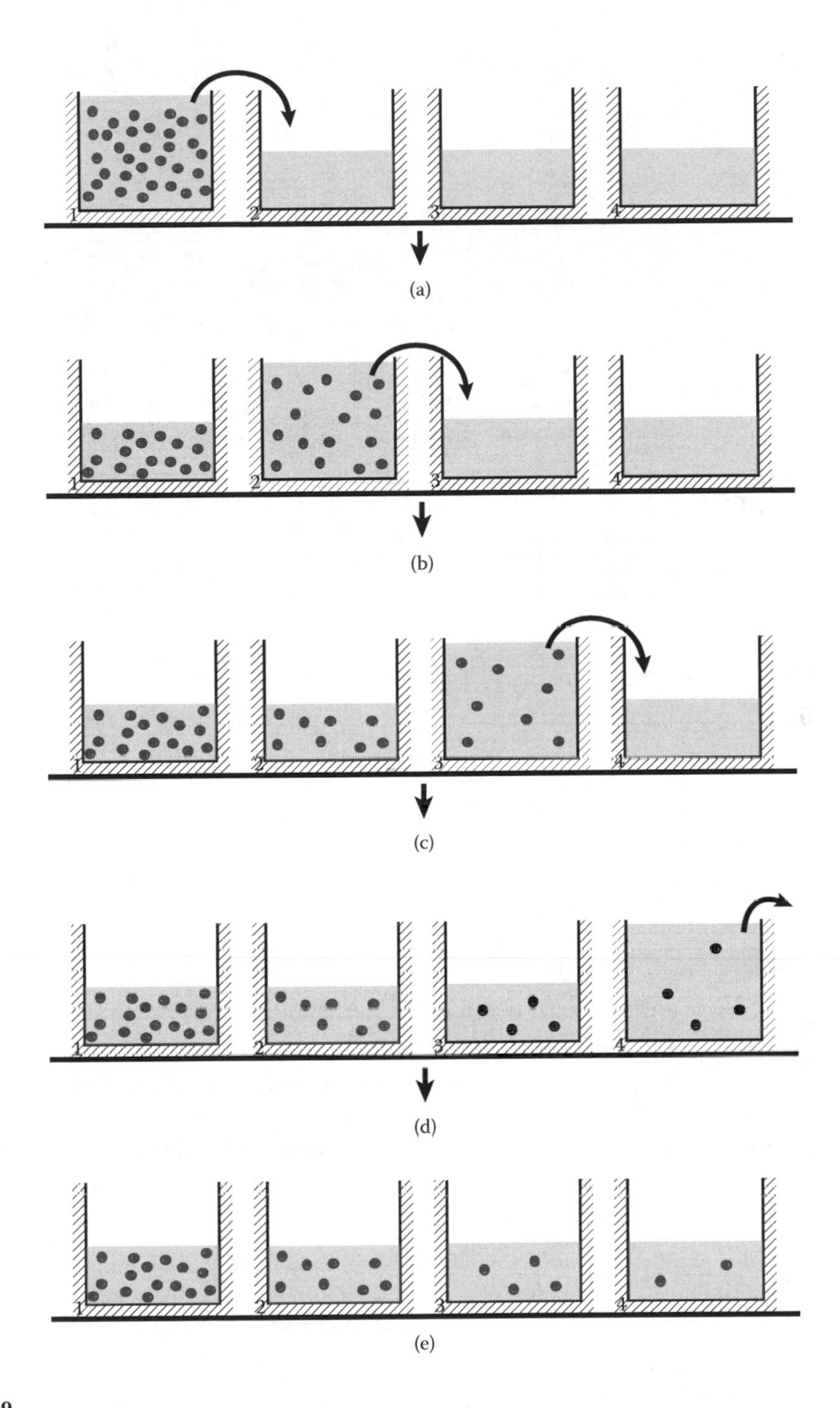

FIGURE 5.9
A cartoon illustration for a one-half serial dilution. (a) 32 cells are added to well #1 in two unit volumes. One unit volume with half of the cells (16 cells) is then transferred to well #2. (b) The 16 cells are mixed with another unit volume of medium in well #2. Afterwards, one unit volume with 8 cells is transferred to well #3. (c) The 8 cells are mixed with another unit volume of medium in well #3. Afterwards, one unit volume with 4 cells is transferred to well #4. (d) The 4 cells are mixed with another unit volume of medium in well #4. Afterwards, one unit volume with 2 cells is discarded. (e) After the serial dilution, wells #1–4 contains one unit volume of medium each and 16, 8, 4, and 2 cells, respectively.

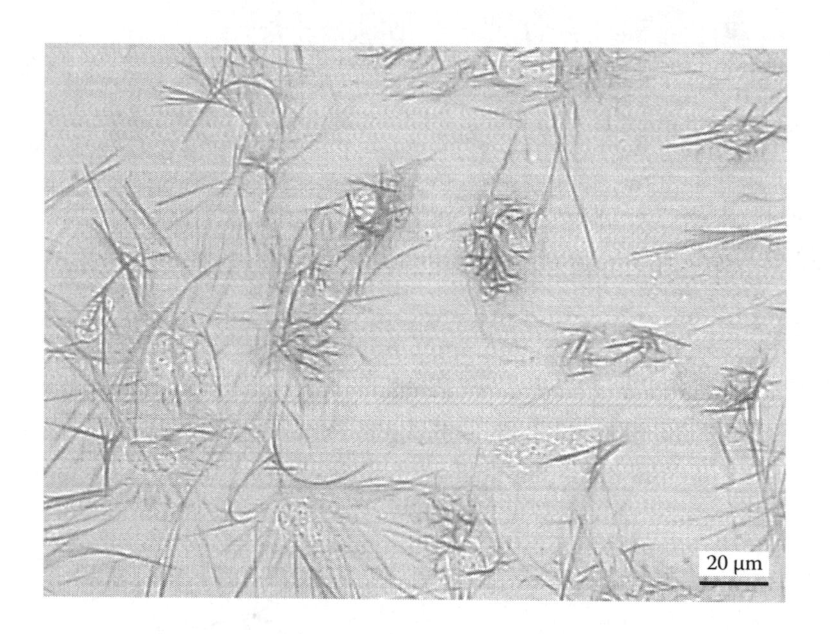

FIGURE 5.10
MTT formazan crystals observed in MC3T3-E1 cells after 2 hours of incubation. The crystals are needle shaped and have a dark purple color. As the crystals grow, the cells will eventually lyse and die but remain attached. In this view, the shapes of many cells are still recognizable. (Image is taken with brightfield at 400× optical magnification.)

- 2.0-ml microcentrifuge tubes
- Disposable plastic cuvettes
- Minicentrifuge
- Spectrophotometer
- Waste container for MTT/formazan

2. *Preparation:* Vent and wipe down the BSC. Aseptically move the items that are labeled with [BSC] in Step 1 into the BSC.
3. *Incubating cells with MTT:* Aspirate the old medium from the 6-well plate with the test materials. Add 1 ml of fresh complete α-MEM back to each well, and then add 100 μl MTT solution to each well using a 100-μl (or 200-μl) micropipette with filter tips. Swirl the medium to mix. Incubate the plate in a 37°C CO_2 incubator for 2 hours. Add MTT to the calibration plate as well.
4. *Observing MTT formazan crystals:* After incubation, observe the cells in the control wells under the microscope. You should be able to see MTT formazan crystals like those shown in Figure 5.10. Take pictures of the crystals for your lab report.
5. *Dissolving the MTT formazan crystals*
   a. Inside the BSC, gently aspirate the medium from all cell cultures. Afterwards, move the plate outside the BSC.
   b. Using a pair of *fine-tipped* jeweler's forceps (Figure 5.11), move the discs of test materials from the original plate (plate #1) to a new plate (plate #2), one disc per well. *Note:* Keep the cell-seeded sides of the discs *up* in the new wells, and try to minimize the contact between the forceps and the discs.

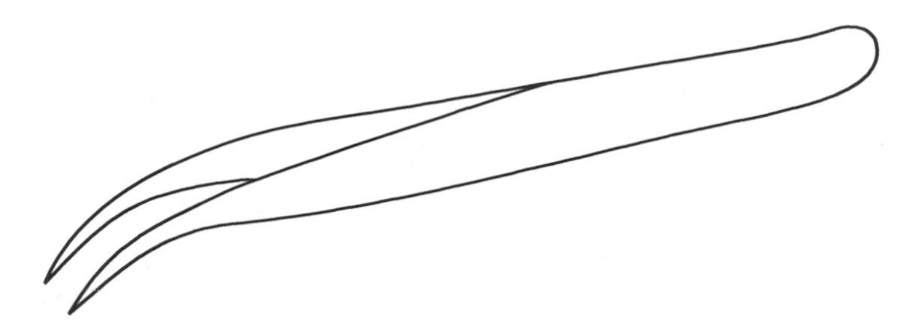

FIGURE 5.11
Jeweler's forceps with curved sharp tips.

c. Add 1500 µl (1000 µl and then 500 µl) of DMSO to each of the two control wells (the ones that did not have discs) in plate #1 and to the wells with discs in plate #2. *Note:* Be patient when pipetting DMSO because it can be quite viscous.

d. Shake the plates 10–20 minutes to dissolve the MTT formazan crystals. Observe the control wells in plate #1 under the microscope from time to time to monitor the dissolution process. Also, observe the color of the solution over time.

e. For the calibration plate, add DMSO to the wells after removing the medium. Shake the plate and check the dissolution progress as described in Step 5d.

6. *Measuring the MTT formazan absorbance*

   a. Transfer the DMSO solutions from the 6-well plates to 2.0 ml microcentrifuge tubes.

   b. Centrifuge the solutions for 3 minutes using a minicentrifuge.

   c. Set the measuring wavelength of the spectrophotometer to 570 nm and measure the absorbance of the supernatant in the 2.0-ml microcentrifuge tubes. Use a new cuvette for each sample, and blank the cuvettes with DMSO. Discard the MTT formazan solution into a designated waste container.

   d. Perform the same procedures for the calibration samples.

7. *Finishing up:* Place the material discs in 5% bleach. (They can be reused after cleaning.) Pipette any solution left in the 2.0-ml tubes, the plates, and the cuvettes into the MTT waste container, and rinse with 70% ethanol if necessary; afterwards, discard these items into the biohazard waste. Store the reagents.

Data Processing

1. *Data from your own group:* Use the calibration curve to determine the (duplicate) cell counts for each test material and the control (polystyrene), and calculate the relative cell counts of the test materials as percentages of the control. (If calibration was not performed, then calculate the relative cell counts using the absorbance measurements.)

2. *Data from class:* Pool the data for the relative cell counts from other groups in the class together with your own, and calculate the average and the standard error for each test material. Use the average relative cell counts to determine how well the test materials support MC3T3-E1 proliferation relative to polystyrene, and use the standard errors to judge the reliability of the data.

Questions

1. You need to seed three T-25 flasks with a mouse fibroblast cell line called NIH 3T3. You need the cultures to be near confluent 3 days later, when the cell count will be 1.0×10^6 in each T-25 flask. For the experiment that you are doing, it is crucial that the three flasks have the same cell count. At the moment you have trypsinized an NIH 3T3 culture and obtained a suspension with a cell count of 5.0×10^5/ml. How would you do the cell seeding? Assume that the doubling time for NIH 3T3 is 24 hours. An optimal volume for culture medium in T-25 flask is 8 ml.
2. For the extraction test, we use a ratio of 2.0 cm^2 surface area to 1.0 ml extraction medium. In standard practices, a commonly used ratio is 6.0 cm^2 surface area to 1.0 ml medium. Given the same extraction time, how does the surface area to extraction medium ratio affect the extraction test?
3. When we perform the direct contact test and the extraction test, we use latex (or gum rubber), a material that is moderately cytotoxic. Why not use a material that is extremely cytotoxic as the positive control to make the results more "striking," so to speak?
4. You are asked to conduct cytotoxicity tests for a new alloy that can potentially be used for biomedical applications. You first conduct a direct contact test by placing slugs of this alloy on top of L-929 cells, and find that the cells directly beneath the samples are all dead, but the cells surrounding the samples appear to be healthy, so you give it a 1 on the cytotoxicity scale. You then conduct an extraction test by extracting the alloy slugs with cell culture medium and incubating L-929 cells with the extraction, and you find that the cell cultures are ≥100% viable compared to the control, so you have to give it a 0 on the cytotoxicity scale. You repeat the two experiments, and find the same results. How do you reconcile the results from the two different tests?
5. Student groups A and B are both conducting an extraction test on a material. The two groups work together to seed two sets of cell cultures with the same seeding density, and they maintain the cell cultures in the same conditions with the same reagents. Days later they conduct the test independently but come up with wildly different results. The two groups are puzzled, and they want to compare notes to see what could be the source(s) of this discrepancy. What would you suggest they look for when they compare notes?
6. Why did we use MC3T3-E1 cells for the experiments in Session 3? Can we use L-929 cells instead?
7. You are running an MTT assay to determine the proliferation of NIH 3T3 cells (see Question 1). Your co-worker happens to be running an MTT assay too, using the L-929 cell line, which as you know is another mouse fibroblast cell line. Thinking about saving some time, you wonder if you could just use your co-worker's calibration curve to calculate the cell counts for your assay. But could you?
8. In the extraction test, we performed cell counting using a hemacytometer with trypan blue staining. So why did we use the MTT assay to count cells when we studied cell proliferation in Session 3? Isn't the hemacytometer method a lot simpler?
9. Teflon is used as a negative control in the experiments in Sessions 1 and 2. Does it serve the same role in the experiments in Session 3?

10. You are asked to test the cytotoxicity of a drug using *in vitro* cell cultures, and you need to test the drug in a concentration range of 0.1 µm to 1 *mM*. You have at your disposal 6-well plates, L-929 cell cultures, and a 10 *mM* stock solution of the drug dissolved in cell culture medium. Sketch a plan to do the testing with answers to the following questions: What test method do you plan to use? How do you set up the cell cultures? How do you make the right drug concentration? How do you obtain and present the experimental results?
11. Propose a plan to test the biocompatibility of a material that you are interested in. Tell us about the chemical and physical properties of this material, why you are interested in testing it, what testing methods you plan to use, and how you are going to use the test results with respect to your interests in this material.

Appendix. Recipes and Sources for Equipment, Reagents, and Supplies*

Session 1. Cytotoxicity Evaluation Using Direct Contact Tests

See BLS II appendix.

Session 2. Cytotoxicity Evaluation Using Liquid Extracts of Materials

See BLS II appendix.

Session 3. Biofunctionality Evaluation through Cell Adhesion/Proliferation Studies

- **Titanium discs.** Cut from titanium sheet (Sigma-Aldrich cat. no. 267503) (can be made by your local machine shop). *Sel. crit.:* Thickness ~0.2 mm, diameter precision 0.1 mm.
- **Stainless steel discs.** McMaster Carr, part no. 2895T53. *Sel. crit.:* Thickness ~0.2 mm, diameter precision 0.1 mm.
- **Teflon discs.** Cut from Teflon sheet (McMaster Carr, part no. 8569K21) (can be made by your local machine shop). *Sel. crit.:* Thickness >0.3 mm, diameter precision 0.1 mm.
- **MC3T3-E1 cells.** MC3T3-E1 Subclone 4, ATCC cat. no. CRL-2593.
- **α-modified minimum essential medium (α-MEM).** Thermo Scientific HyClone α-MEM, mfr. no. SH30265.01, Fisher cat. no. SH3026501. *Sel. crit.:* Same or similar compositions.
- **Complete α-MEM (100 ml).** 88 ml α-MEM, 10 ml FBS, 1 ml 100X pen/strep stock solution (see BLS II appendix (Chapter 4), Section II.6) and 1 ml 100X sodium pyruvate stock solution (Thermo Scientific HyClone sodium pyruvate solution, mfr. no. SH30239.01, Fisher cat. no. SH3023901).
- **MTT.** Sigma-Aldrich, cat. no. M5655. *Sel. crit.:* For cell culture use.

* *Disclaimer*: Commercial sources for reagents listed are used as examples only. The listing does not represent endorsement by the author. Similar or comparable reagents can be purchased from other commercial sources.

- **DMSO.** Sigma-Aldrich, cat. no. D2438. *Sel. crit.:* For cell culture use.
- **Fine-tipped jeweler's forceps.** Fisherbrand dissecting jeweler's microforceps, Fisher cat. no. 08-953F. *Sel. crit.:* With sharp tips.
- **2.0-ml microcentrifuge tubes.** Fisher cat. no. 05-408-138. *Sel. crit.:* Most fit mini centrifuges.
- **Disposable plastic cuvettes.** See Module I (Chapter 2) appendix (Session 2).
- **Spectrophotometer.** See Module I (Chapter 2) appendix (Session 2).

References

FDA, Use of international standard ISO-10993, Biological evaluation of medical devices, Part 1: Evaluation and testing, *General Program Memorandum G95-1*, GPO, Washington, D.C.,1995.

ISO, Biological evaluation of medical devices, Part 5: Tests for *in vitro* cytotoxicity, *ISO 10993-5*, 1999.

Kirkpatrick, C., F. Bittinger, M. Wagner, H. Köhler, T. van Kooten, C. Klein, and M. Otto, Current trends in biocompatibility testing, *Proc. Inst. Mech. Eng. [H]*, 212, 75–84, 1998.

Mosmann, T., Rapid colorimetric assay for cellular growth and survival: Application to proliferation and cytotoxicity assays, *J. Immunol. Methods*, 65, 55–63, 1983.

Williams, D.F., ed., Definitions in biomaterials. *Proceedings of a Consensus Conference of the European Society for Biomaterials, Chester, England, 1986*, Elsevier, Amsterdam, 1987.

6

Module IV. Tissue Engineering: Organotypic Culture of Skin Equivalent

There has always been the need to repair, replenish, or replace tissues and organs as the human body suffers from diseases, injuries, or simply undergoes aging. Clinically, such needs have been met with medicinal treatments, implants, and organ transplants, with various degrees of success. However, medicinal treatments often have their limits; for example, it is known that for some injuries such as thrashed muscle, if the damage is beyond a certain threshold, the tissue or organ simply will not heal. Implants have been saving lives and improving the quality of life for many patients; devices such as pacemakers, heart valves, stents, and hip-joint replacements are standard treatments at present and will probably remain so for the foreseeable future. Organ transplants were a major medical advancement, the results of which range from completely restored health to at least a prolonged life. However, the availability of organs is a major issue, as are immunological reactions caused by complications. In the quest for solutions to such problems, a new area—regenerative medicine—has emerged as a result of rapid advancement in molecular and cell biology, physiology, chemistry, material sciences, and other relevant disciplines, and one of its main disciplines is tissue engineering. The goal of tissue engineering is "the development of functional tissues and organs *in vitro* for implantation *in vivo* or for direct remodeling and regeneration of tissue *in vivo* to repair, replace, preserve or enhance tissue or organ function lost due to disease, injury, or aging" (NIH 2006). Tremendous progress has been made in the past couple of decades in tissue engineering; we can now engineer bone, cartilage, blood vessels, skin, and cornea (Figure 6.1), and we can look forward to tissue-engineered liver, heart, pancreas, and other organs that will repair or replace the failed ones in patients. Another exciting development is in stem cell research. Stem cells are cells that can perpetually renew themselves but can also differentiate into multiple cell types. For example, it has been shown that mesenchymal stem cells found in our bone marrow can differentiate into osteoblasts (bone-forming cells), chondrocytes (cartilage-forming cells), myocytes (heart muscle cells), adipocytes (fat cells), and even neuronal cells. A major breakthrough took place in 2007 when Dr. Yamanaka and co-workers discovered four genes that can reprogram somatic cells such as skin cells to resemble embryonic stem cells, which opens the door to tissue engineering of organs that are genetically the patients' own.

A well established approach to growing tissues or organs *in vitro* is the use of scaffolds, in which cells and materials are assembled to mimic the arrangement of cells and the extracellular matrix (ECM) in tissues and organs; these "live" scaffolds are then cultured *in vitro* for eventual implantation. This principle is illustrated in Figure 6.2, using tissue engineering of skin as an example. Materials are a critical consideration for this approach. Naturally, biocompatibility of materials is a first criterion, for which both the biosafety and the biofunctionality aspects must be given important considerations. Biodegradability is also required since a desired outcome for an implanted tissue or organ is the eventual integration with its *in vivo* environment; thus, biodegradable natural and synthetic polymers

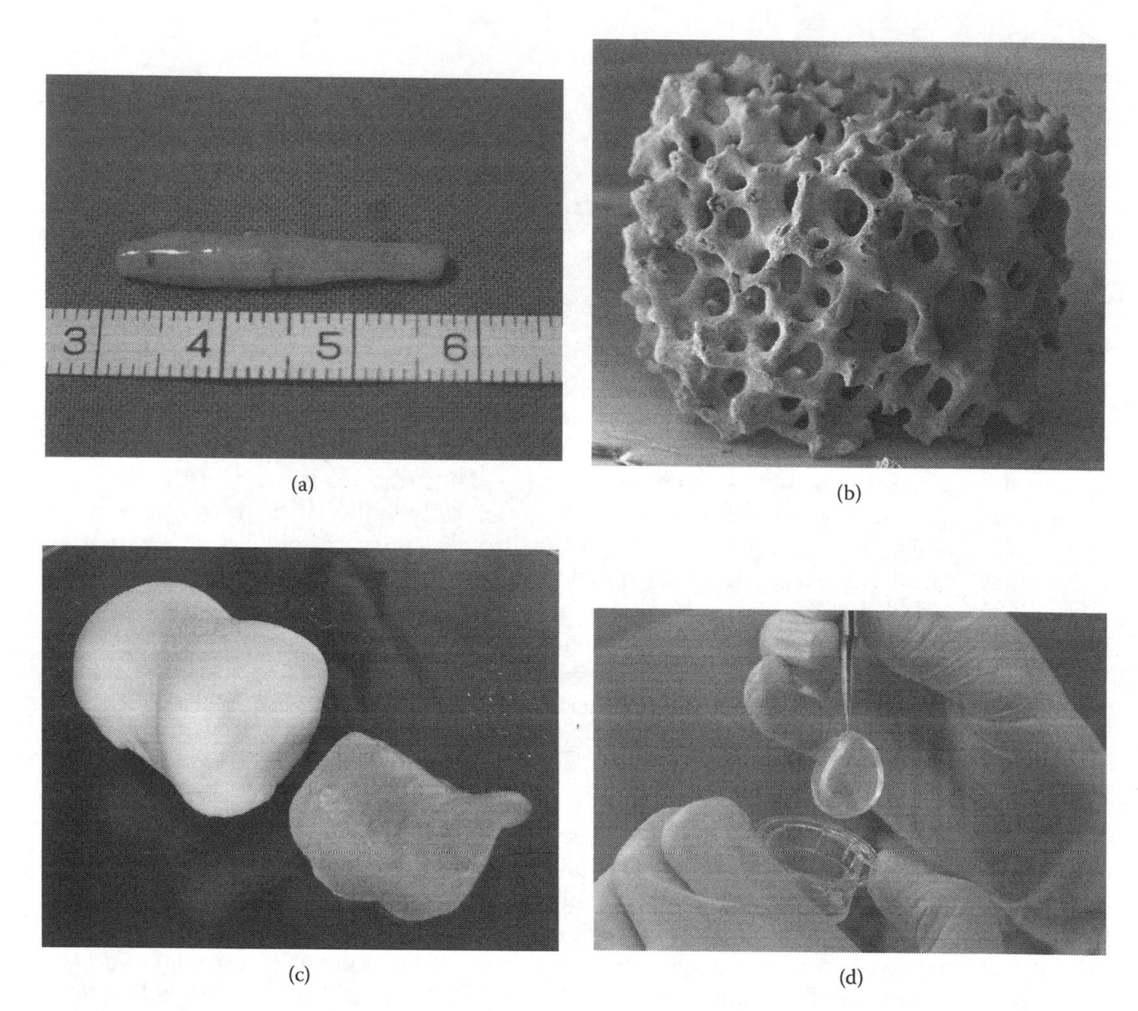

FIGURE 6.1
Tissue engineering examples. (a) A segment of blood vessel is constructed using smooth muscle cells and endothelial cells differentiated from mouse embryonic stem cells. Reprinted by permission from Macmillan Publishers (Shen et al. 2003). (b) A porous bioceramics scaffold that allows bone ingrowth after implantation. The BD 3D Calcium Phosphate Scaffold is a proprietary mineralized calcium phosphate bioceramic that is ideal for *in vitro* and *in vivo* analyses of bone metabolism and cartilage regeneration. Image copyrighted, courtesy Becton, Dickinson, and Co. (c) A breakthrough self-assembly technique for growing replacement cartilage is used to grow the entire articular surface that is tailored three-dimensionally to fit a specific lower femur (Hu and Athanasiou 2006). Image courtesy of J. Hu and K. A. Athanasiou, Dept. of Bioengineering, Rice University. (d) Tissue-engineered cornea is constructed using a hydrogel scaffold and epithelial, keratocytes, and endothelial cells (Griffith et al. 1999). Photo courtesy of the National Research Council of Canada.

and resorbable bioceramics are widely used in tissue engineering. Additional material properties are required for specific applications. For example, mechanical strength is often important for materials that are used for constructing bone-replacement tissues, but soft materials are much more suitable for engineering neural tissues. Moreover, encapsulation and release of factors such as growth factors and enzymes is becoming an important aspect for the *in vitro* development of tissues or organs (Figure 6.2), and accommodating these factors adds another consideration when choosing the right materials.

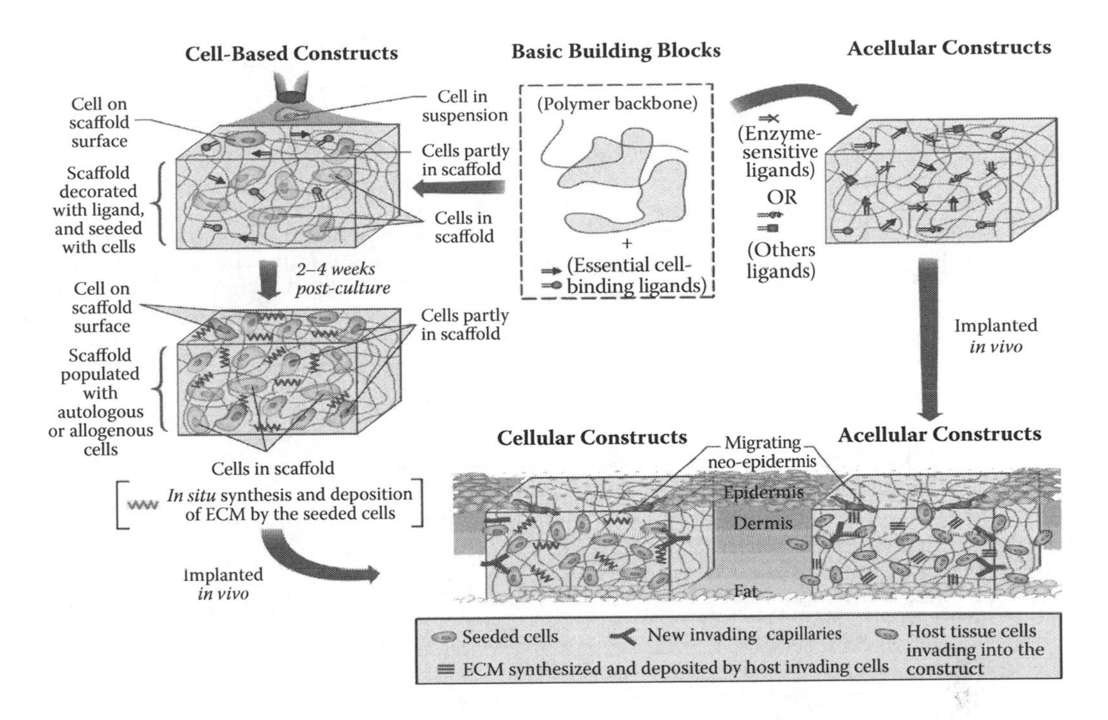

FIGURE 6.2
The basic building blocks of a tissue-engineered construct are a biopolymer, one or more biomimetics, and perhaps cells. The biopolymer must be biocompatible and biodegradable. Biomimetics are selected to add function to the biopolymer. The bioactive function may be cell-binding activity, growth factor activity or growth factor-binding activity, or enzymatic activity or enzyme-binding activity. Cells added exogenously to the engineered biopolymer may be used to induce a functioning tissue substitute for transplantation (left schemata). These cells may be stem cells or genetically engineered cells. When an acellular biopolymer is implanted, enough information must be available within the engineered construct to support endogenous tissue cell ingrowth and appropriate differentiation for tissue formation (right schemata). Reprinted by permission from Macmillan Publishers (Clark et al. 2007).

In this module we will explore organotypic tissue engineering of skin. Skin is composed of two main tissues: the epidermis and the dermis (Figure 6.3). In the dermis, type I collagen is the major component of the ECM, and fibroblast cells are responsible for secreting collagen as well as a number of growth factors that stimulate the growth of the epidermis. The epidermis is formed and renewed by keratinocytes, a type of epithelial cells that makes keratin, a tough material that comprises the outer surfaces of our bodies. Currently a number of tissue-engineered products are available as skin-replacement materials (MacNeil 2007); among them, the epidermal/dermal replacement materials such as Apligraf are tissue-engineered using an organotypic approach. "Organotypic" is defined as resembling an organ *in vivo* in three-dimensional form or function, or both. Organotypic skin-replacement products have been successfully used for clinical treatment of wounds and ulcers. Although the results of commercialization of these types of products have been mixed (Bouchie 2002), the potential for tissue-engineered, "living" skin replacement is still enormous, especially with the fact that it is now possible to tissue engineer skin replacement with a patient's own cells cultured from hair follicles (Limat et al. 2003; Tausche et al. 2003). In addition to clinical use, organotypic skin equivalents have been routinely used as skin models for product testing and research (MacNeil 2007).

Epithelial Cover

Involves the delivery of autologous keratinocytes as one of the following:

- An integrated sheet such as Epicel (Genzyme Tissue Repair). This is developed from the methodology originally pioneered in 1981 (O'Connor et al. 1981). A biopsy of the patient's cells is grown into an integrated sheet and enzymatically detached for delivery to the patient (Wright et al. 1998).
- Subconfluent cells on a carrier such as Myskin (CellTran) (Haddow et al. 2003). Cells are delivered to the patient before they reach confluence on a chemically defined carrier dressing.
- Small sheets cultured from a patient's hair follicles such as Epidex (Modex Therapeutics) (Tausche et al. 2003).
- A spray such as CellSpray (Clinical Cell Culture). Subconfluent cells are expanded in the laboratory and made into a suspension in which they are transported. They are then delivered to the patient as a spray (Navarro et al. 2000).

Dermal replacement materials

- Donor skin: skin from screened skin donors can be used to provide either a temporary wound cover or a permanent source of allodermis (Hermans 1989).
- Integra (Integra LifeSciences): an alternative to donor skin that provides a vascularized dermis for a subsequent split-thickness skin graft (Stern et al. 1990).
- Alloderm (Lifecell): freeze-dried human donor dermis (Wainwright et al. 1996).
- Dermagraft (Advanced Biohealing): a synthetic material conditioned with donor fibroblasts (Marston et al. 2003).
- Transcyte (Advanced Biohealing): similar to Dermagraft but with a silicone membrane to act as a temporary epidermal barrier (Kumar et al. 2004).
- Permacol (Tissue Science Laboratories): porcine skin that provides a temporary wound dressing (Jarman-Smith et al. 2004).

Epidermal/dermal replacement materials

- Apligraf (Organogenesis): combines allogeneic keratinocytes and fibroblasts with bovine collagen to provide a temporary skin-replacement material suitable for use in chronic wounds but not major burns (Bello and Falabella 2003).
- Orcel (Ortec International): combines allogeneic keratinocytes and fibroblasts with bovine collagen to provide a temporary skin-replacement material suitable for use in chronic wounds (Lipkin et al. 2003).
- Cincinnati skin substitute, or Permaderm (Cambrex): comprises autologous keratinocytes and fibroblasts crafted into reconstructed skin with bovine collagen. Can provide a permanent skin substitute for burns patients (Boyce et al. 2006).

In the following experiments, we will grow an organotypic, "living" skin-replacement material (skin equivalent) *in vitro.* Two essential components are needed for an organotypic culture of skin equivalent: the dermis equivalent, which consists of a collagen matrix with fibroblast cells at the minimum, and the epidermis equivalent, which consists of layers of keratinocytes. In addition, melanocytes (pigment cells) and dendritic cells (immunological cells) have been successfully incorporated into skin equivalents. There are a number of methods for organotypic culture of skin equivalent, most of which are based on the principle illustrated in Figure 6.2. We will use this principle to create "bare bones" but functioning organotypic skin equivalents: we will create the dermis equivalent by encapsulating fibroblasts in a matrix of collagen (the natural biomaterial that we studied in Module II), and then we will construct the epidermis equivalent by growing keratinocytes on top of the collagen matrix. Stimulants (ascorbic acid and growth factor) will be added to the cultures to help promote the proper development of both layers. The organotypic culture process will finish with stratification of the uppermost layers of the epidermis equivalent.

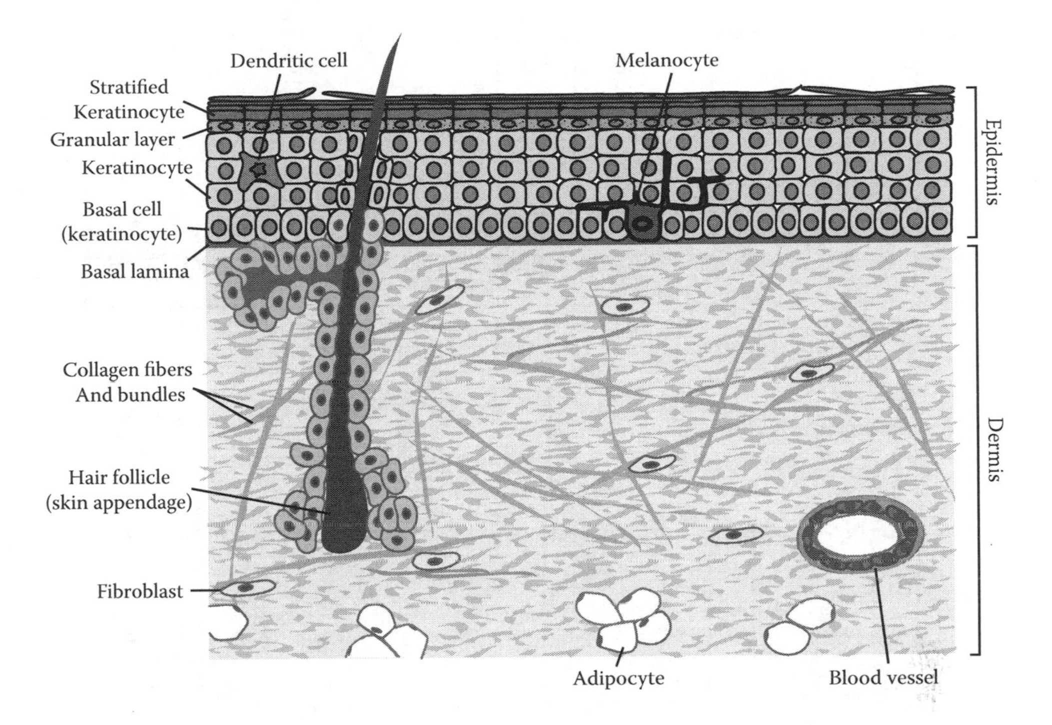

FIGURE 6.3
Structure of skin, the largest organ in the human body. Skin can be divided into two principle layers, the dermis and the epidermis. The epidermis consists of layers of keratinocytes in different stages of differentiation; the stratified outermost layers provide protection from mechanical damages while the granular layer seals out water. Other cell types such as melanocyte and dendritic cells help to protect the skin from UV and infection. The basal lamina defines the border between the epidermis and the dermis. The main component of the dermis is an ECM rich in collagen and elastin, which are secreted by fibroblasts. Adipocytes can be found in the deep portion of the dermis—the hypodermis. The dermis also contains a network of blood vessels and sensory nerves (not depicted). Skin appendages include hair follicles and sweat glands (not depicted) that are distributed in various part of the skin.

Session 1. The Dermis Equivalent

In this session, we will construct the dermis equivalent by encapsulating fibroblast cells in a collagen matrix. The "trick" for the encapsulation is based on a unique property of type I collagen: The tropocollagen molecules self-assemble at 37°C to form a solid gel but not at cold temperatures (2–8°C). In practice, sterilized type I collagen (such as the kind that we extracted from calf skin in Module II) is first dissolved in an acidic solution; when the solution is neutralized to pH ~7 at cold temperatures, the tropocollagen molecules will remain soluble. At this point, fibroblast cells are mixed with the collagen solution, and the cell-collagen mixture is subsequently incubated at 37°C, during which the collagen forms a gel with encapsulated cells (Figure 6.4). Note that most cells, including fibroblasts, can survive cold temperatures for a period of time. On the other hand, acidic pH is lethal for most cells; therefore, it is critical for the acidic collagen solution to be properly neutralized. (This tends to be the most difficult part of the experiment since the solution must remain *cold* throughout the titration process.)

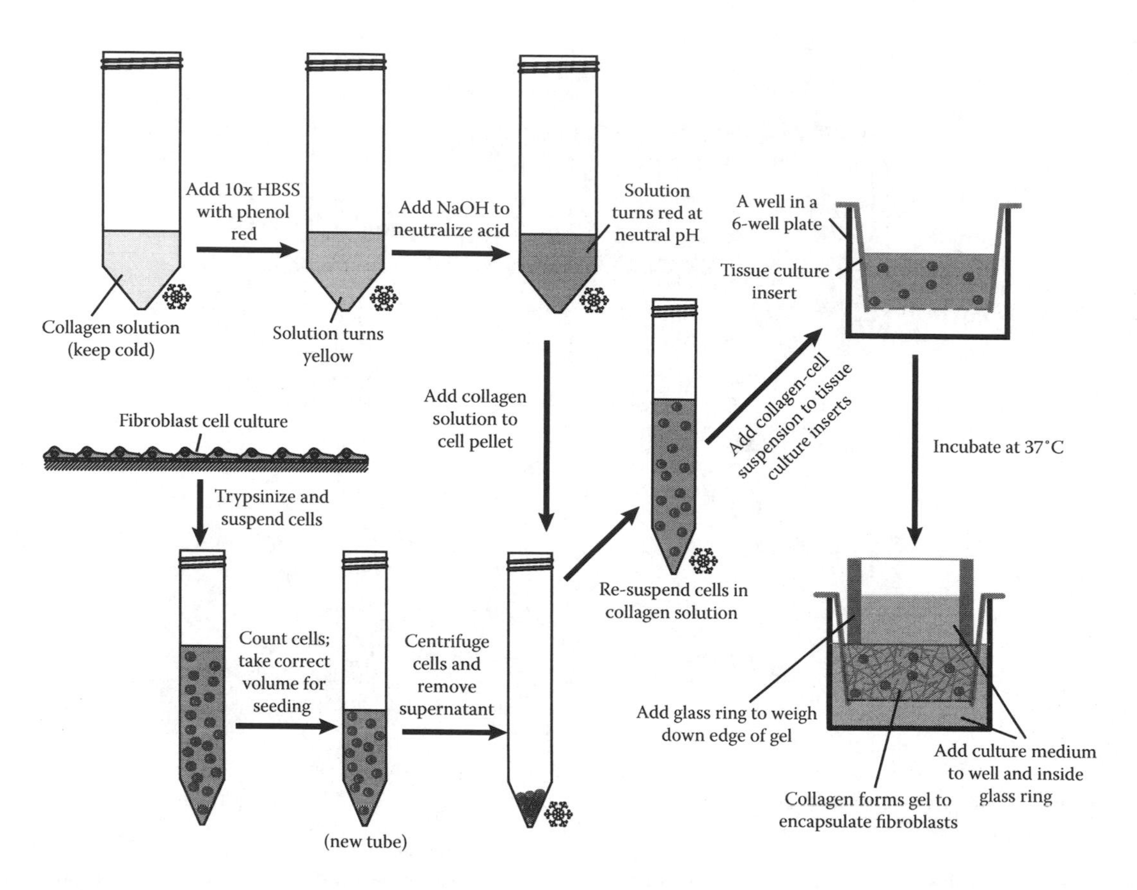

FIGURE 6.4
Constructing the dermis equivalent. An acidic solution of collagen is first neutralized with NaOH while being kept cold. An NIH 3T3 cell culture is then trypsinized, suspended, and counted, and a volume of the cell suspension is used based on the target cell count for the encapsulation. The cells are gently pelleted by centrifugation, chilled, and then resuspended in the cold, neutralized collagen solution. The collagen–cell suspension is then dispensed into tissue culture inserts (Figure 6.5), which serve as molds for collagen gel formation. The collagen solution forms solid gel during incubation at 37°C, encapsulating the cells. Afterwards, a glass ring is added on top of the collagen gel to weigh down its perimeter, the purpose of which is to counter the contraction forces exerted by the cells. Cell culture medium is then added to begin the culturing process. The dermis equivalent will be cultured for >1 week before the addition of the epidermis equivalent (Figure 6.7).

We will use a mouse skin fibroblast cell line, NIH 3T3—a frequently used "model" fibroblast cell line, as the resident cells in the dermis equivalent. Most commercially available skin-equivalent products, either for clinical applications or for research, use primary cultures of fibroblasts to mimic the physiological properties of natural skin as closely as possible. However, since primary human cell cultures are subjected to stringent regulation (and expensive to culture), we will use a cell line instead. NIH 3T3 is a spontaneously transformed cell line that retains most of the behaviors of its physiological counterpart, one of which is secretion of collagen, and the other is the rearrangement of the ECM through exertion of forces by the cells. After encapsulation, the NIH 3T3 cells contract the collagen gel as they pull on the collagen fibers in the matrix; therefore, a counterforce—the weight of the glass ring—is applied to limit the contraction along the z-axis only (Figure 6.4). When cultured in the presence of ascorbic acid, the NIH 3T3 cells are induced to secret collagen (Russell et al. 1981), which helps to enhance the collagen matrix of the dermis equivalent (see Figure 6.2). Furthermore, ascorbic acid plays a critical role in the hydroxylation of the proline and lysine residues in collagen, a key step in the cross-linking of collagen, which in turn is key to the mechanical strengthening of the collagen matrix. (This is also why ascorbic acid, a.k.a. vitamin C, is the cure for scurvy, a disease for which deficiency of collagen cross-linking is the main cause.)

Procedures

Part 1 (Day 1). Encapsulation of NIH 3T3 Cells in Collagen Gel

1. *Check-in*
   a. Samples and materials
      - 5–10 mg/ml collagen* in 0.1 *M* acetic acid solution, chilled (see Appendix A)
      - NIH 3T3 cell culture
   b. Reagents
      - 10X Hank's balanced salt solution (HBSS) with phenol red, chilled
      - [BSC] PBS
      - [BSC] Complete Dulbecco's Modified Eagle's Medium (DMEM), with 10% FBS, 4 g/L glucose, 2 *mM* L-glutamine, 1 *mM* sodium pyruvate, 50 IU/ml penicillin and 50 μg/ml streptomycin, warmed up to 37°C
      - [BSC] Trypsin EDTA, warmed up to room temperature
      - 2.0 *M* NaOH, chilled
      - 1.0 *M* HCl, sterile filtered, chilled
      - Trypan blue
   c. Special equipment and supplies
      - [BSC] 6-Well plate, sterile
      - [BSC] Two tissue culture inserts (Figure 6.5), sterile
      - [BSC] Tweezers, autoclaved
      - [BSC] Two glass rings, autoclaved
      - [BSC] Two 15-ml conical centrifuge tube, sterile
      - [BSC] One 50-ml conical centrifuge tube, sterile
      - Ice
      - Hemacytometer
2. *Preparation*
   a. Turn on the blower of the BSC at least 30 minutes ahead of time, and then wipe it down with 70% ethanol. Afterwards, aseptically move the items that are labeled with [BSC] into the BSC.
   b. Open the tissue culture inserts, and using a pair of sterile tweezers, place the inserts into the 6-well plate. Label the plate with your group's name and the date.

* The concentration of collagen solution does not need to be precise as long as it is able to form gel. On the other hand, the more collagen used, the thicker the collagen matrix will be, and thus the stronger the dermis equivalent.

FIGURE 6.5
Tissue culture insert. A tissue culture insert is a cuplike vessel that is placed inside a multiwell microplate to create two compartments in the plate well. Its bottom is a porous membrane that allows the exchange of culture medium but not cells; thus, the two compartments can host two different cultures and allow exchange of molecules between them. (In our experiments, tissue culture inserts are used first as molds for culturing the dermis equivalent and then as devices for air exposure of the epidermis equivalent.) Tissue culture inserts should always be handled with a pair of tweezers and not by hand to avoid contamination.

c. Chill the collagen solution, the 10X HBSS, the NaOH solution, and the HCl solution with ice and keep them cold until use. (Since the ice is not sterile, it is best to keep the ice bucket outside of the BSC.)

3. *Neutralizing the collagen solution*

   a. Inside the BSC pipette 4.5 ml collagen solution into a 50-ml conical centrifuge tube, and then mix in 0.5 ml 10X HBSS. Note the color of the solution. Chill the 50-ml tube with collagen solution on ice for about 5 minutes.

   b. *Note:* A successful titration of the acidic collagen solution with NaOH is critical for this experiment. Remember to:

      – Keep the collagen solution cold because neutralized collagen will spontaneously gel at room temperature. Chill the solution on ice every 2–3 minutes, even if it means moving it back and forth between the ice and the BSC; do make sure that the tube containing the collagen solution is capped before bringing it out of the BSC and wiped it with 70% alcohol before bringing it in. Alternatively, you can use an autoclaved 500-ml beaker half-filled with ice inside the BSC, but remember that the ice is nonsterile, so do not let the ice or the melted water touch the neck of the tube, and wipe the tube with 70% ethanol frequently.

      – The neutralization reaction between NaOH and acetic acid will generate heat; therefore, it's crucial to add NaOH in tiny amounts and mix it with the collagen solution as quickly as possible. (One trick you can try is to "dot" a drop of NaOH solution on the wall of the tube, and then in several rapid swirls of the collagen solution, the NaOH solution is quickly diluted.)

   c. Titrate the collagen solution with 2.0 *M* NaOH solution. (Ask your instructor for an approximate volume so you have an idea about how much NaOH is needed, if possible.) The correct color should be orange-red (such as the color of the complete DMEM). If the color is orange-yellow, add 5 µl NaOH solution

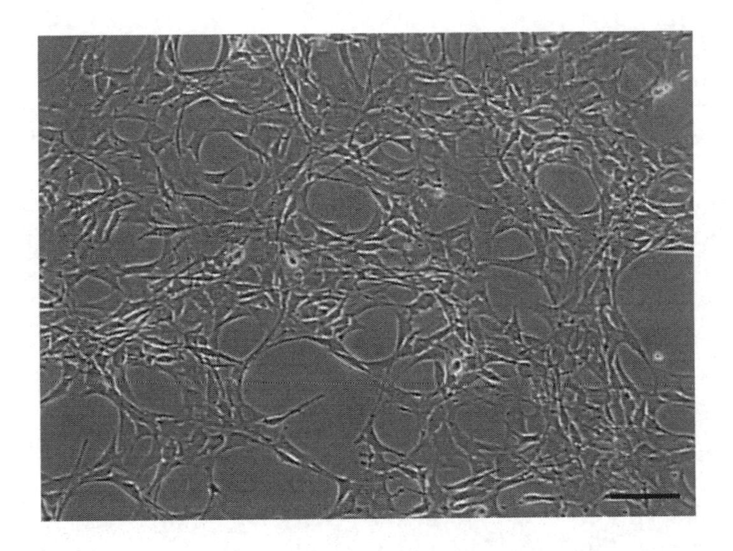

FIGURE 6.6
NIH 3T3 cells. Note that these cells have the typical fibroblast morphology with spread, elongated cell bodies and extensions called lamellipodia and filopodia. Also note that the cells maintain extensive contacts with other cells. Magnification: 100. Scale bar: 100 μm. Image is obtained using an inverted optical microscope set to phase contrast.

each time until the color changes to orange-red. If the color becomes purplish-pink/fuchsia, you have overshot; back-titrate with 1.0 *M* HCl. *Important:* Keep the collagen on ice after the titration.

4. *Trypsinizing, suspending, and counting NIH 3T3 cells*
   a. Observe the NIH 3T3 cell culture. Note the morphology of the cells (Figure 6.6) and the confluency of the culture. Use standard procedures (see Chapter 4) outlined below to trypsinize, suspend, and count the cells:
   b. Rinse the T-25 flask with 5 ml of PBS. Aspirate the PBS.
   c. Add 0.5 ml trypsin EDTA, and incubate for 2 minutes.
   d. Suspend the cells with 4.5 ml complete DMEM.
   e. Take out a small amount of the cell suspension and keep the rest on ice.
   f. Mix 20 μl of the cell suspension with 20 μl of trypan blue.
   g. Load the cell-dye mixture onto a hemacytometer, and count cells in 10 squares.
   h. Calculate the cell count:

$$\text{C.C.}\left(\text{ml}^{-1}\right) = N \times 10^3 \times 2$$

   with *N* the total number of cells in 10 squares. Clean the hemacytometer.

5. *Prepare NIH 3T3 cells for seeding*
   a. The target number of cells to seed is 4×10^5 cells (for two samples). Calculate the volume of cell suspension needed:

$$V_{cs} = \frac{4 \times 10^5}{\text{C.C.}} \text{ (ml)}$$

b. Transfer V_{cs} ml of the cell suspension into another 15-ml tube, and centrifuge it at 1000 ×g for 3 minutes. Remove the supernatant and chill the tube on ice for 1 minute. (If $V_{cs} > 5$ ml, that means you do not have enough cells, which is OK; just use all of the cell suspension and centrifuge it in the original tube. Make a note of the actual seeding number in your lab notebook.)

6. *Encapsulating the cells in collagen*
   a. After both the neutralized collagen and the cell pellet are sufficiently chilled, add the collagen solution to the cell pellet, and gently pipette up and down to resuspend the cells.
   b. You should have ~5 ml of collagen/cell suspension. Pipette ~2.5 ml into each tissue culture insert in the 6-well plate. Avoid foaming. Immediately move the plate to a 37°C CO_2 incubator. Incubate for 30 minutes or longer.
7. *Setting up the dermis equivalent cultures*
   a. Check the gelation status of the collagen by tilting or shaking the plate slightly. If the collagen solution has formed a gel, it should not flow. Note the color and transparency of the gel. Move the plate into the BSC.
   b. Add a sterilized glass ring on top of each collagen gel using a pair of tweezers. Do not press down. Aspirate any liquid that is "squeezed out" from the gel.
   c. Add 2 ml of complete DMEM to the well, and 1 ml inside the tissue culture insert for each culture.
   d. Move the 6-well plate to a 37°C CO_2 incubator. Culture the dermis equivalents for at least 1 week with a change of medium every 2–3 days (see Part 2).
8. *Finishing up:* Discard the 15-ml and 50-ml conical tubes into the biohazard waste, dispose of the trypan blue solution waste properly, store the reagents, and wipe down the BSC with 70% ethanol.

Part 2 (Day 2, etc.). Adding Ascorbic Acid and Maintaining the Cultures

This part can be performed by the instructor if the students' scheduling is an issue.

1. *Check-in*
   a. Samples and materials
      - The dermis equivalent cultures
   b. Reagents
      - Complete DMEM supplemented with 50 μg/ml ascorbic acid
2. *Adding ascorbic acid:* Two to three days after the encapsulation, remove the old medium from the dermis-equivalent cultures; replenish with ascorbic-acid-supplemented complete DMEM.
3. *Culture maintenance:* Change medium every 2–3 days with ascorbic-acid-supplemented complete DMEM. Maintain the cultures for at least 1 week.

Session 2. The Epidermis Equivalent

The epidermis of skin is consisted mainly of keratinocytes, with the presence of several other cell types such as melanocytes in relatively very small numbers. In natural skin,

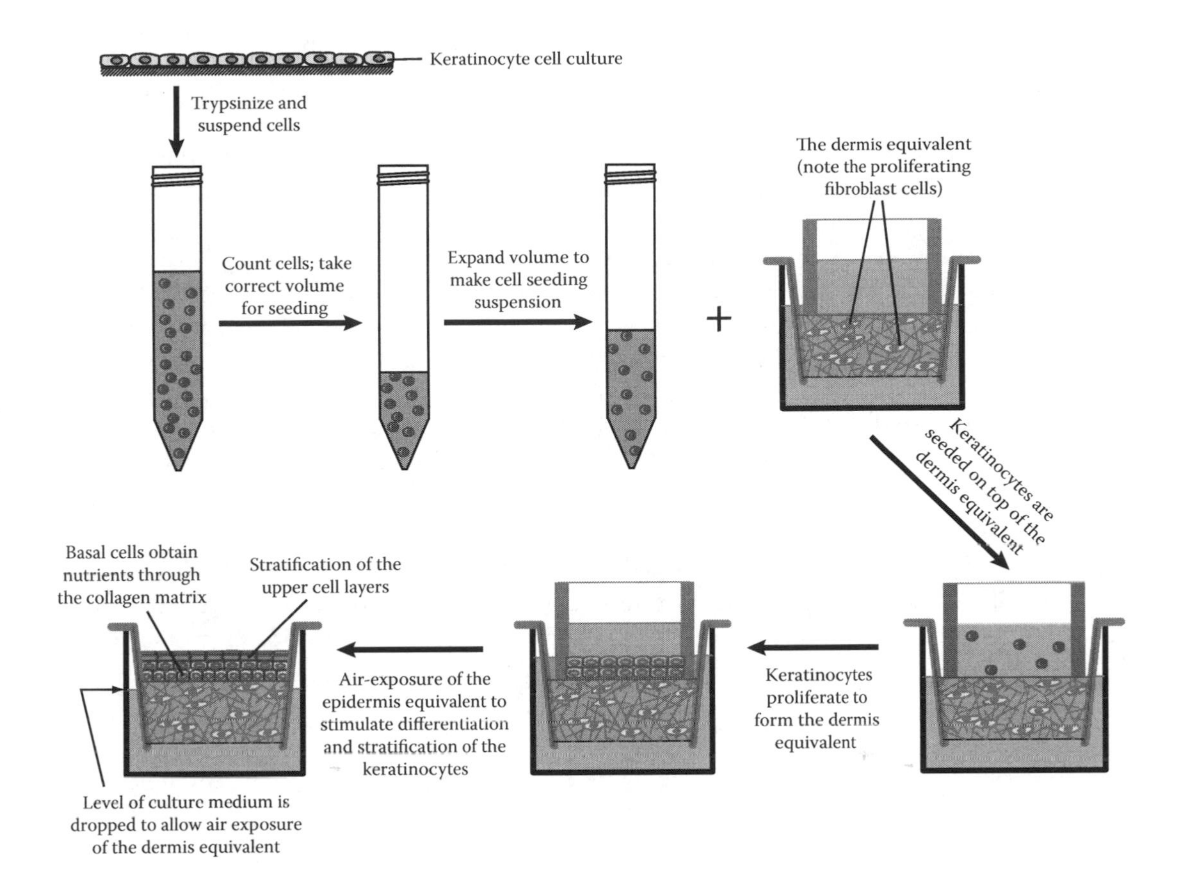

FIGURE 6.7
Constructing the epidermis equivalent. Keratinocyte cells are seeded on top of the dermis equivalent (Figure 6.4), which should be well established with proliferating fibroblasts at this point. The glass ring helps to contain the cells for the seeding process. Stimulated by growth factors that are secreted by the fibroblasts in the dermis equivalent, the keratinocytes proliferate and form a multilayered epidermis equivalent. The upper layers of cells of the epidermis equivalent undergo stratification upon air exposure.

keratinocytes differentiate to form several distinct layers (see Figure 6.3). In tissue-engineered skin equivalent, the epidermis equivalent can be constructed using keratinocytes only. The keratinocytes are seeded on top of the dermis equivalent and cultured for a period of time, during which the keratinocytes are stimulated by the growth factors secreted by the fibroblasts inside the dermis equivalent and subsequently proliferate and grow into multiple layers with distinct cell morphologies. This is in contrast with the homogeneous monolayer that they form when cultured on 2D plastic surfaces. In the meantime, a basal lamina is developed by the keratinocytes at the dermis-epidermis boundary as well. To culture fully developed organotypic skin equivalents, the outmost layers of epidermis equivalents need to be stratified. The "trick," or the switch that turns on the stratification process, is air exposure. In practice, this means reducing the level of the culture medium to just below the epidermis layers (Figure 6.7). The reduced moisture prompts the upper layers of the epidermis to stratify, forming corny layers that are similar to the stratum corneum of natural skin.

For commercial skin-equivalent products, again, in order to mimic the physiological properties of natural skin as closely as possible, primary cultures of keratinocytes are used. An excellent source for human keratinocytes is neonatal (baby) foreskin from circumcision in hospitals due to its consistency in quality and availability and the ability of the cells to proliferate vigorously. Again, for simplicity we will use a keratinocyte cell line, HaCaT. HaCaT cells are spontaneously immortalized keratinocytes derived from human adult skin; these cells retain important functions of their native counterparts such as differentiation, stratification, and keratin synthesis. In skin-equivalent cultures, HaCaT cells can form multiple layers, differentiate, and stratify, but with some degrees of deficiency: Proliferation is delayed, more fibroblast cells are required, and the upper layer is thinner and not fully stratified. These deficiencies are due to the genetic alterations that have taken place during the transformation of HaCaT that result in low production levels of several critical factors (Stark et al. 2004). These deficiencies can be significantly compensated by transforming growth factor-α (TGF-α), a growth factor that plays critical roles in the signaling pathways of keratinocytes (Maas-Szabowski et al. 2003; Pastore et al. 2007).

In this session, we will seed HaCaT cells on top of the dermis-equivalent cultures from Session 1, and allow them to proliferate for at least 1 week (or better yet, 2 weeks, if possible). To study the effect of TGF-α, we will culture one of the skin equivalents in the presence of TGF-α and the other in its absence. To complete the organotypic culturing process, we will air-expose the epidermis equivalent in order to trigger the stratification of its upper layers. Finally, we will prepare the two cultures for histological analyses in Session 3.

Procedures

Part 1 (Day 1). Seeding HaCaT Cells onto the Dermis Equivalent

1. *Check-in*
   a. Samples and materials
      - Cultures of NIH 3T3 in collagen matrices (from Session 1)
      - HaCaT culture in T-25 flask
   b. Reagents
      - [BSC] PBS with 0.05% EDTA, sterile filtered
      - [BSC] Complete DMEM supplemented with 50 μg/ml ascorbic acid, sterile, warmed up to 37°C
      - [BSC] Trypsin EDTA, warmed up to room temperature
      - Trypan blue
   c. Special equipment and supplies
      - [BSC] One 15-ml conical centrifuge tube, sterile
      - Hemacytometer
2. *Preparation*
   a. Vent the BSC and wipe it down with 70% ethanol. Afterwards, aseptically move the items labeled with [BSC] in Step 1 into the BSC.
   b. Observe the morphology of HaCaT cells (Figure 6.8) under the microscope and note the confluency of the culture. Pay attention to the difference in morphology between HaCaT and the NIH 3T3 cells (see Figure 6.6).

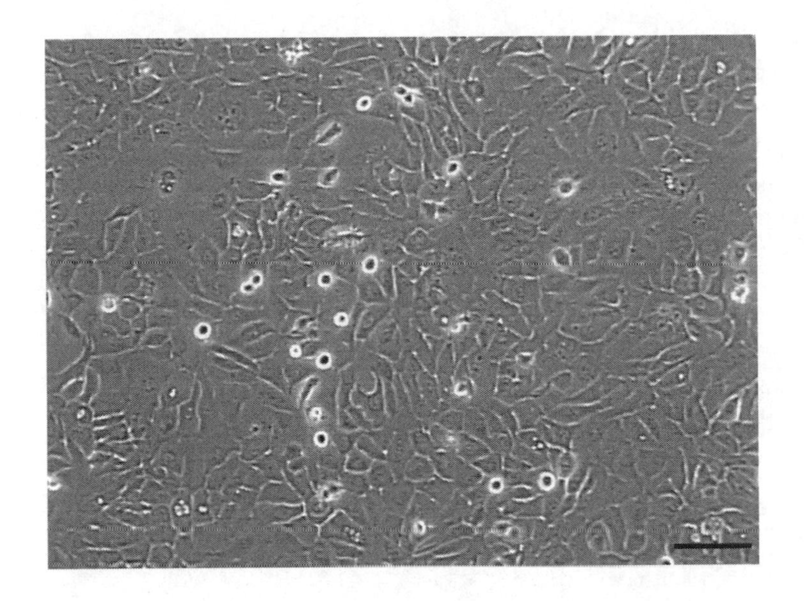

FIGURE 6.8
HaCaT cells. These cells tend to be cuboidal in shape and have tight cell-to-cell contacts that result in the formation of cobblestone-like cell clusters. Magnification: 100. Scale bar: 100 μm. Image is obtained using an inverted optical microscope set to phase contrast.

3. *Trypsinizing and suspending HaCaT cells*
   a. *Note:* Procedures for trypsinizing HaCaT cells are slightly different from the standard ones that we have been using for fibroblasts or osteoblasts.
   b. Aspirate the medium, and then rinse the cells with 4 ml of PBS with 0.05% EDTA. Aspirate the rinse, and again add 4 ml of PBS with 0.05% EDTA. Incubate the cells in a 37°C CO_2 incubator for 5 minutes. *Note:* The purpose of this step is to break down the intercellular connections by removing Ca^{2+} through EDTA chelating; the cells will remain attached to the flask.
   c. After the incubation, aspirate the PBS/EDTA, and add 0.5 ml of trypsin EDTA to cover the cells. Incubate at 37°C for 1 minute.
   d. Add 4.5 ml of complete DMEM to suspend the cells. Store the cell suspension on ice.
4. *Counting cells:* Use standard procedures outlined in Session 1, Step 4.
5. *Observing the dermis equivalents:* In the BSC, inspect the dermis equivalents and note their colors and textures. Do not remove the glass rings. With the tip of a Pasteur pipette, very lightly touch the collagen matrix to test its resiliency.
6. *Seeding HaCaT cells*
   a. The target number of cells to seed is 4×10^5 cells (for two samples), calculate the volume of HaCaT cell suspension needed:

$$V_{cs} = \frac{4 \times 10^5}{\text{C.C.}} \text{ (ml)}$$

   with C.C. as the cell count from Step 4.

b. If V_{cs} <2 ml, add the V_{cs} ml of the cell suspension into a 15-ml tube, and then add complete DMEM for a total volume of 4 ml.

c. Remove the old medium from the dermis-equivalent cultures.

d. Add 0.5 V_{cs} ml of the original HaCaT suspension (from Step 3) or 2 ml of the diluted suspension (from Step 6b) inside each glass ring. (If V_{cs} >5 ml, you do not enough cells; then just use all the cells by adding 2.5 ml of the cell suspension into each insert. Be sure to make a note of the actual seeding number in your lab notebook.)

e. Move the cultures to a 37°C CO_2 incubator and incubate for ~24 hours.

7. *Finishing up:* Discard the 15-ml tube in biohazard waste, dispose of the trypan blue solution waste properly, store the reagents, and wipe down the BSC with 70% ethanol.

Part 2 (Day 2, etc.). Adding TGF-α and Maintaining the Cultures

This part can be performed by the instructor if the students' scheduling is an issue.

1. *Check-in*
   a. Samples and materials
      - Skin-equivalent cultures
   b. Reagents
      - Complete DMEM supplemented with 50 µg/ml ascorbic acid, warmed up to 37°C
      - Complete DMEM supplemented with 50 µg/ml ascorbic acid and 2 ng/ml TGF-α, warmed up to 37°C
   c. Special equipment and supplies
      - Tweezers, autoclaved
2. *Removing the glass rings:* Two or three days after seeding the HaCaT cells, remove the glass rings using a pair of tweezers. Be careful if the glass ring has attached to the collagen matrix, in which case you can use the tip of a Pasteur pipette to gently hold down the collagen sheet while lifting the ring.
3. *Adding TGF-α:* Label one culture as "+TGF-α" and the other as "–TGF-α" on the lid. Remove the old medium from the cultures. Add complete DMEM with or without TGF-α to the corresponding wells.
4. *Culture maintenance:* Change the medium every 2–3 days for both cultures using the corresponding media. (Make sure that the two cultures and media are not mixed up.) Maintain the cultures for 2 weeks if possible, or 1 week at the very least.

Part 3 (Day 3). Air Exposure of the Epidermis Equivalent

1. *Check-in*
   a. Samples and materials
      - [BSC] Skin-equivalent cultures
   b. Reagents

- [BSC] Complete DMEM with 50 µg/ml ascorbic acid, warmed up to 37°C
- [BSC] Complete DMEM with 50 µg/ml ascorbic acid and 2 ng/ml TGF-α, warmed up to 37°C

c. Special equipment and supplies
- [BSC] Tweezers, autoclaved
- [BSC] (Optional) Two custom-made Teflon supports in polypropylene (PP) containers, autoclaved (see Appendix B)

2. *Preparation:* Vent and wipe down the BSC. Aseptically move the items labeled with [BSC] in Step 1 into the BSC. Observe the cultures and note their colors, textures, and shapes (any wrinkles or rolled-up edges?).
3. *Air exposure of the epidermis equivalents in a 6-well plate*
   a. For both cultures, aspirate the medium from the wells, and *pipette* (do not aspirate) from inside the inserts.
   b. Add each medium to the corresponding *well* to just below the tissue culture insert: take up ~1.5 ml medium in a serological pipette; hold its tip against the wall of the well and deliver slowly while looking down through the collagen matrix; watch carefully when the liquid make first contact with the bottom of the insert, and stop once the liquid has made *full* contact. Make sure that there are no bubbles trapped underneath the inserts or the cultures may dry out partially (it's better to over-fill the well than to under-fill it); also make sure that there is no liquid on the surface of the cultures. Ask your instructor to check and see if the air exposure is OK.
   c. Move the cultures to a 37°C CO_2 incubator. Since the amount of medium in each well is small, a medium change needs to be performed every day, or whenever the color of the medium becomes yellowish. Allow 1 week or at least 3 days of air-exposure time.
4. *(Alternative to step 3) Air exposure using Teflon supports in PP containers* (see Appendix B)
   a. Label the PP containers as "+TGF-α" and "–TGF-α." With a pair of tweezers, place the inserts into the Teflon supports in the corresponding PP containers.
   b. Add each medium to the corresponding PP container to just below the tissue culture insert: take up ~6 ml medium in a serological pipette; hold its tip against the wall of the container and deliver slowly while looking down through the culture; watch carefully when the medium makes first contact with the bottom of the insert, and stop once full contact is made. Make sure that there are no bubbles trapped underneath the inserts (it's better to over-fill the well than to under-fill it), and that there is no liquid on the surface of the cultures. Ask your instructor to check and see if the air exposure is OK.
   c. Cap the containers, but loosen the caps by about half a turn (use a point on the cap as reference). Move the containers to a 37°C CO_2 incubator. The medium needs to be changed every 2 or 3 days, or whenever the color of the medium becomes yellowish. Allow 1 week or at least 3 days of air-exposure time.
5. *Finishing up:* Dispose of the used items properly, store the media, and wipe down the BSC with 70% ethanol.

Part 4 (Day 4). Fixing the Skin-Equivalent Samples for Histology Slide Preparation

1. *Check in*
   a. Samples and materials
      - [BSC] Skin-equivalent cultures
   b. Reagents
      - Fixation solution (3.7% formaldehyde in PBS)
      - [BSC] PBS
   c. Special equipment and supplies
      - Tweezers
      - Digital camera
      - Waste Container for the fixation solution
2. *Observation*
   a. Remove the medium from the wells (or the PP containers) for both cultures. Rinse the inside of each insert with 2 ml of PBS. (As usual, do not aspirate from inside the inserts.)
   b. Move the cultures out of the BSC. Observe the gross appearance of the cultures again and note their colors, textures, and shapes. Use a pair of tweezers to lightly tug at the edges of the cultures to test their mechanical strength, and push the tip of the tweezers lightly into the matrix to test their resiliency. How does the elasticity of the collagen matrix compare to that of a rubber band? What does the surface of the cultures look like?
   c. Use a digital camera to take pictures of the cultures.
3. *Fixing the cultures with formaldehyde*
   a. *Note:* The fixation process is essentially the extensive covalent cross-linking of the proteins in the sample by formaldehyde (Figure 6.9). Such cross-linking preserves the structure and organization of the tissue.

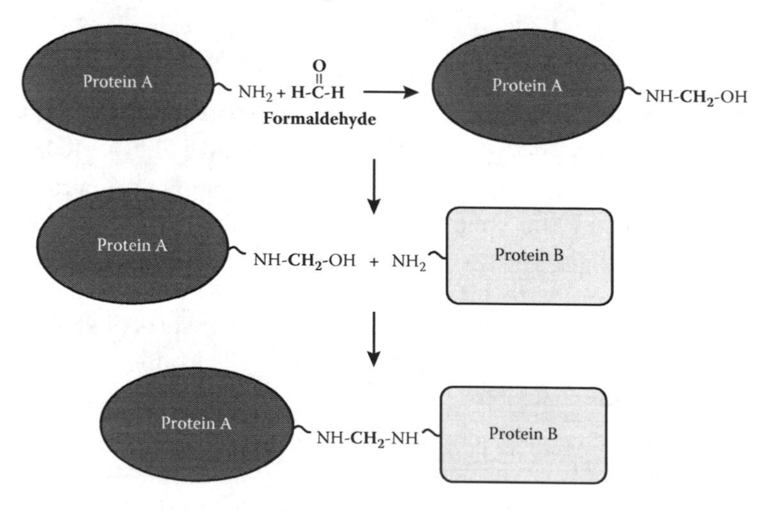

FIGURE 6.9
Protein cross-linking by formaldehyde. Formaldehyde reacts mostly with amine groups on amino acid side chains that are on protein surfaces. Each formaldehyde molecule can potentially form covalent bonds with two amine groups from two different proteins, A and B, thus cross-linking the two proteins.

b. Add a sufficient amount of fixation solution to submerge the skin-equivalent cultures. Cover the samples and allow the fixation to proceed for 10 minutes at room temperature.

c. Cut a piece of Parafilm about 5 cm × 5 cm in size. After 10 minutes of fixation, pick up one of the skin equivalents with a pair of tweezers and lay it down on the Parafilm. Using a razor blade, cut a ~0.5 cm strip near the middle of the skin-equivalent culture, so that representative thin sections can be cut from either side (Figure 6.10). Using a *pencil* (do *not* use pens or markers), label an embedding cassette (Figure 6.11) with the name of the sample. Make sure that you keep track of the "+TGF-α" and "–TGF-α" samples in the sample

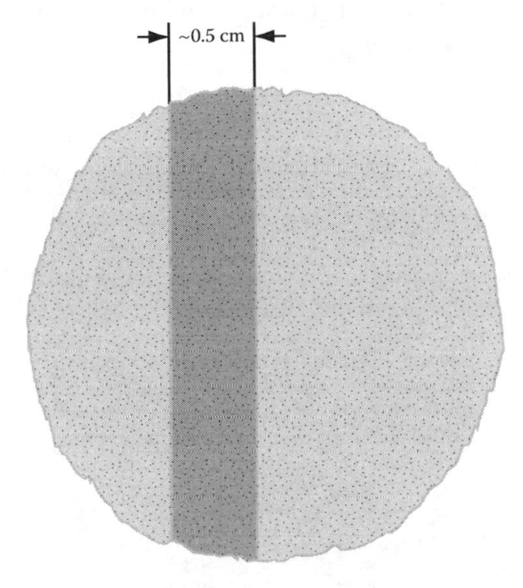

FIGURE 6.10
Dissecting out a sample of an organotypic skin equivalent for histology slide preparation.

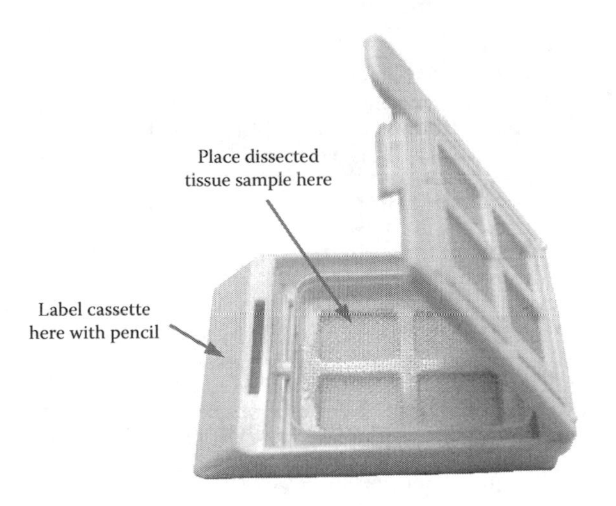

FIGURE 6.11
Embedding cassette that is used for histology slide preparation.

names. Place the cutout strip of the skin equivalent in the cassette, lock the cassette, and place it in a container provided by your instructor. The cassettes should be submerged in fixation solution. Process the other culture using the same procedures.

4. *Preparing histology slides:* You instructor will send the fixed skin-equivalent samples to a histology facility for slide preparation.
5. *Finishing up:* Dispose of the leftover fixation solution in a designated waste container. Discard the tissue culture inserts and the 6-well plate. Discard the leftover skin equivalents in the biohazard waste container (or preserve them in the fixation solution). If Teflon supports and PP containers were used, leave them in 5% bleach for >10 minutes before washing, rinsing, and hanging them up to dry. (They are reusable.)

Preparing Histology Slide

A histology slide is a very thin, and usually stained, cross-section of a tissue. We can learn much about a tissue by observing histology slides of the tissue under a microscope. Histology slide preparation is now largely automated, but it still requires a lot of effort and skill. Many universities and hospitals have core facilities for histology to which tissue samples can be submitted. What happens to your skin-equivalent samples once they are submitted to one of these histology facilities? The following is an outline of a typical slide preparation process.

- Fixation: Tissues are fixed in lab, and after dissection, samples of the tissues are kept in formaldehyde solution throughout the submission process. The samples will be further fixed for a period of time (usually overnight) in the histology facility before processing.
- Dehydration: Eventually the water in the tissue samples need to be replaced with wax (paraffin) so that the tissue samples can be cut into thin sections. The dehydration process removes water by soaking the tissue samples in a series of alcohol-water mixtures with increasing alcohol concentration until 100% alcohol is reached.
- Clearing: Since alcohol is not miscible with paraffin, it needs to be "cleared out" by another solvent, xylene, which is miscible with both alcohol and paraffin. The tissue samples are soaked in several changes of xylene for the clearing.
- Infiltrating: The dehydrated and cleared tissue samples are soaked in xylene with increasing amounts of paraffin until the samples are in pure paraffin, thus the samples are "infiltrated" with paraffin. The temperature needs to be kept above the melting point of paraffin for this process.
- Embedding: The tissue samples are embedded manually in a block of paraffin in the right orientation, so that the appropriate cross-section can be cut out. (For our skin-equivalent samples, we need cross-sections that show both the epidermis and the dermis equivalent. In histotechnical terms, we need the samples to be embedded "on edge.")
- Sectioning: This step requires the fine skills of the histotechnicians. A block of paraffin with embedded tissue sample is fixed onto a microtome—an instrument for thin sectioning—and a sharp blade is used to cut thin (5–10 μm) slices from the block. The cross-sections, which are basically thin wax films, are floated on warm water, which helps to spread and smooth out the films. Afterwards, specially treated glass slides are used to pick up the cross-sections. Since the surfaces of the treated glass slides are positively charged, and proteins in tissues are overall negatively charged, this electrostatic interaction helps to adhere the cross-sections to the slides.
- Staining: Before staining, the tissue samples need to be rehydrated by using the clearing and dehydration reagents in reverse order. There are a number of staining methods, and one of the more commonly used ones is hematoxylin and eosin (H&E): hematoxylin prefers to bind to negatively charged molecules such as DNA and stain them blue or purple, and eosin prefers to bind to positively charged molecules and stain them pink. The staining process can be fully automated.
- Slide mounting: The slides are again dehydrated with alcohol and cleared with xylene, which is allowed to evaporate and leave behind completely dried and stained tissue cross-sections. The final step is to preserve the sample cross-sections on the slides by applying mounting medium in liquid form and placing a coverslip on top of each slide. The mounting medium solidifies to form an optically clear layer of resin, permanently encasing the cross-section between the slide and the coverslip.

Session 3. Histological Studies of Natural Skin and Organotypic Skin Equivalent

Histology is the study of the structures of tissues at the microscopic resolution, which spans a range from hundreds of microns to submicrons for light microscopy or further down to nanometers when electron microscopy is employed. Histological studies reveal a variety of information about tissue composition, organization, interaction, and so forth. In tissue engineering, histology and related studies such as immunohistochemistry and cytochemistry are routinely used to evaluate experimental results.

One of the goals of this module is for us to appreciate the complexity of tissues and see the obstacles that need to be overcome in tissue engineering. In this session, we will first conduct a histological study of natural human skin, in which we observe the architecture of skin as well as its various features. We will then study the organotypic skin equivalents that we have cultured and seek answers to the following questions:

1. In what ways do the skin equivalents resemble natural skin, and in what ways do they not?
2. Comparing the two skin equivalents, does the presence of TGF-α have any effect? If so, what kinds of effects?
3. What are the possibilities and obstacles in tissue engineering of skin replacement?

Procedures

Part 1 (Day 1). Histology of Human Skin (Thick Skin)

1. *Check-in*
   a. Samples and materials
      - A histology slide of human musculoskeletal skin (thick skin)
   b. Special equipment and supplies
      - A slide adaptor for the mechanical stage (if necessary)
2. *Using the microscope*
   a. If you are using a phase contrast microscope, adjust the phase slider to bright-field, and place the slide on the slide adaptor. It is OK to use either upright or inverted microscope. Either way, it would be best if you can move the slide using a mechanical stage.
   b. When observing the slide, start with the lowest magnification and move to higher magnification to focus on particular features. *Observe the features from the eyepieces first* before taking pictures with the camera. Why? (The view from the eyepieces is not limited by the digital resolution of the camera.)
   c. If the view from the eyepieces is dim, you should 1) increase the light intensity when increasing the magnification, and 2) make sure that the light path is not split between the eyepieces and the camera.

3. *The epidermis:* Observe the following epidermal features (Figure 6.12) in the slide. Take pictures of the identifiable features.
   a. The cornified layer (*stratum corneum*): This flaky layer is formed by dead cells that have lost their cytoplasm and consists of mostly lipids and keratin (Figures 12A and 12B).
   b. Sweat ducts: These are helical passages through the epidermis. The cross-sections of these helical structures usually appear as stacks of pores (Figures 12A and 12B, white arrows).
   c. The granular cell layer (*stratum granulosum*): These cells are characterized by dark granules that mark the initial stage of stratification of the cells (Figure 12C, white arrow).
   d. The prickle cell layer (*stratum spinosum*): These cells are cuboidal in shape, and have prickly cell-cell connections, hence the name "prickle cells" (Figure 12D, white arrow).
   e. The basal cell layer (*stratum germinativum*): This is the layer of the cells that borders the dermis. Notice the columnar morphology of the cells (Figure 12D, white arrow head).
   f. The basal lamina: This is a thin layer that borders the dermis. This feature is generally difficult to capture in image but may be visible under the microscope (Figure 12D, black arrow).
4. *The dermis:* Identify the following dermal features in the slide. Again, take pictures of the identifiable features.
   a. Collagen: The extensive fibrous matrix in the dermis is largely collagen. Bundles of collagen fibers are identifiable. Collagen stains light pink in H&E samples (Figure 13A, black arrows).

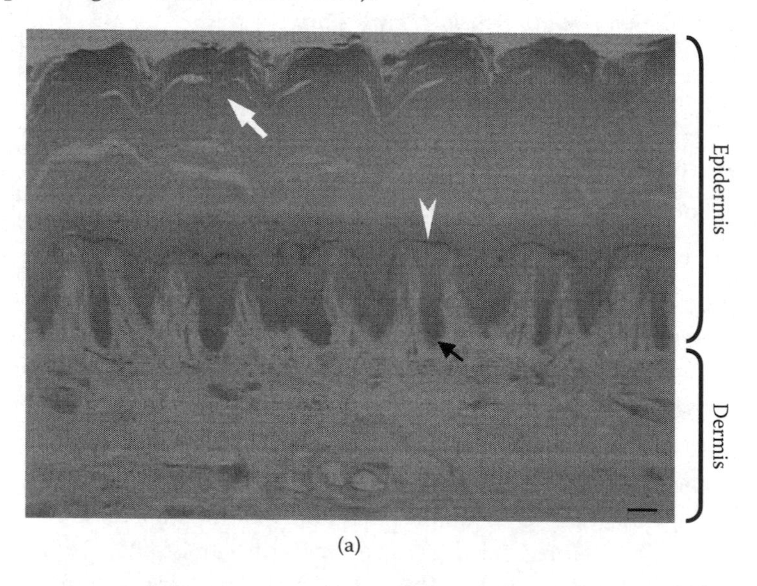

(a)

FIGURE 6.12
Histological features of the epidermis of (thick) skin. (a) An overview of the epidermis relative to the dermis. A sweat duct (white arrows) can be seen in the stratified layer of the epidermis. A thin layer of dark, granular cells (white arrow head) defines the border between the stratified layer and the prickle cells below. The basal cell layer (black arrow) is at the border between the epidermis and the dermis. Magnification: 40. Scale bar: 100 μm.

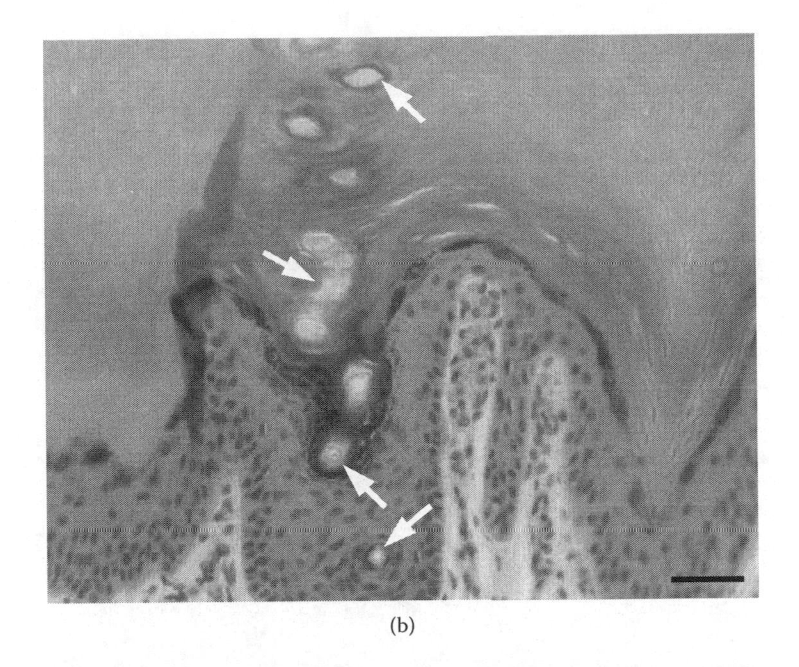

(b)

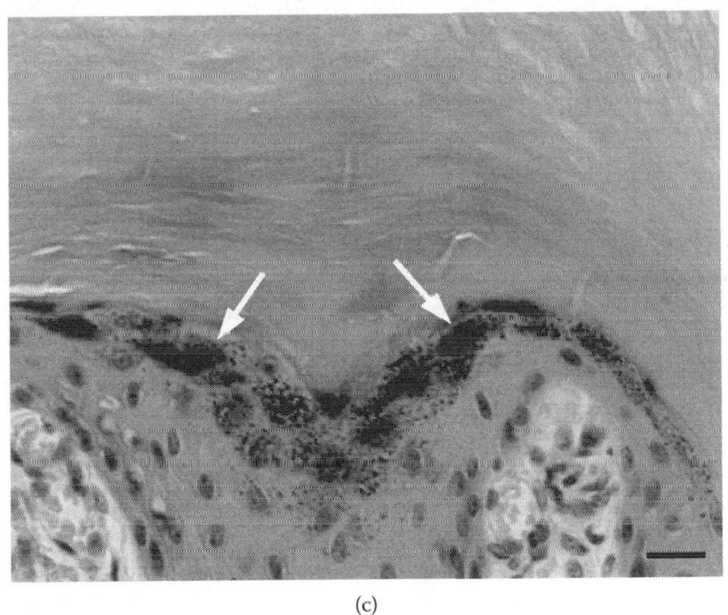

(c)

FIGURE 6.12 (continued)
(b) A sweat duct (white arrows) "snakes" through the epidermis. Magnification: 200. Scale bar: 50 μm. (c) Granular cells with characteristic dark granules (white arrows). Magnification: 400. Scale bar: 20 μm.

b. Fibroblasts: Only the darkly stained nuclei of fibroblasts are usually visible, and they are usually located near collagen fibers (Figure 13A, black arrow heads). Notice the sparse distribution of these cells in the dermis.
c. Adipocytes: These are fat cells that are found in the hypodermis. Adipocytes are derived from fibroblasts, but they have become thousands of times bigger than

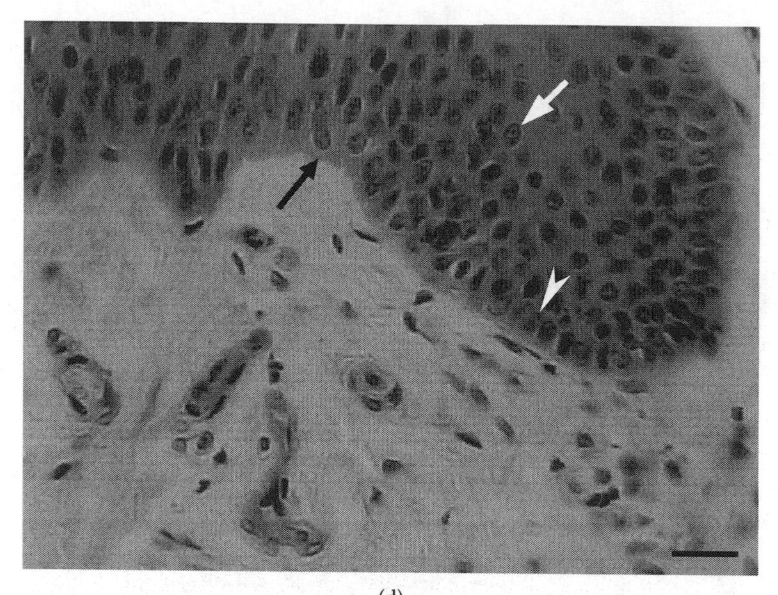

(d)

FIGURE 6.12 (continued)
(d) Prickle cells (white arrow), basal cells (white arrow head), and the location of the basal lamina (black arrow). (The basal lamina is not very visible in this image.) Magnification: 400. Scale bar: 20 µm. All images are obtained using an inverted optical microscope set to brightfield.)

fibroblasts in volume. Notice the large bodies of these cells (Figure 13B, black arrow); fat droplets are stored in the cell bodies while their nuclei, which appear as dark spots, are "squished" against the cell walls (Figure 13B, black arrow head).

d. Blood vessels: Typically, blood vessels appear as a circular arrangement of cells with a lumen. (Sometimes disc-shaped red blood cells remain in the lumen.) Arteries (Figure 14A, white arrow) have thick walls with multiple layers of smooth muscle cells. Veins have thin walls with fewer layers of cells compared to arteries (Figure 14B, white arrow), and their shapes are sometimes distorted during sample preparation due to the relatively weak mechanical strength of the walls.

e. Eccrine sweat glands and ducts: Typical eccrine sweat gland appears as a circular cluster of cells (Figure 15A, white arrows) with a lumen in the middle (Figure 15A, white arrow heads), and the ducts are the tubular structures (Figure 15B, black arrow) that extend from the glands toward the surface of the skin (see Figure 12B). Eccrine sweat glands are part of the body's temperature control mechanisms.

f. Dermal papillae and Meissner's corpuscles: Dermal papillae are the fingerlike extensions of the dermis into the epidermis (Figure 16A). At the tip of some of the dermal papillae are Meissner's corpuscles (Figure 16B), nerve endings that detect light touches on the skin.

g. Pacinian corpuscles: The function of Pacinian corpuscles is to detect heavy pressure on the skin. These nerve structures are usually found in the hypodermis and have the layered appearance of a dissected onion (Figure 6.17).

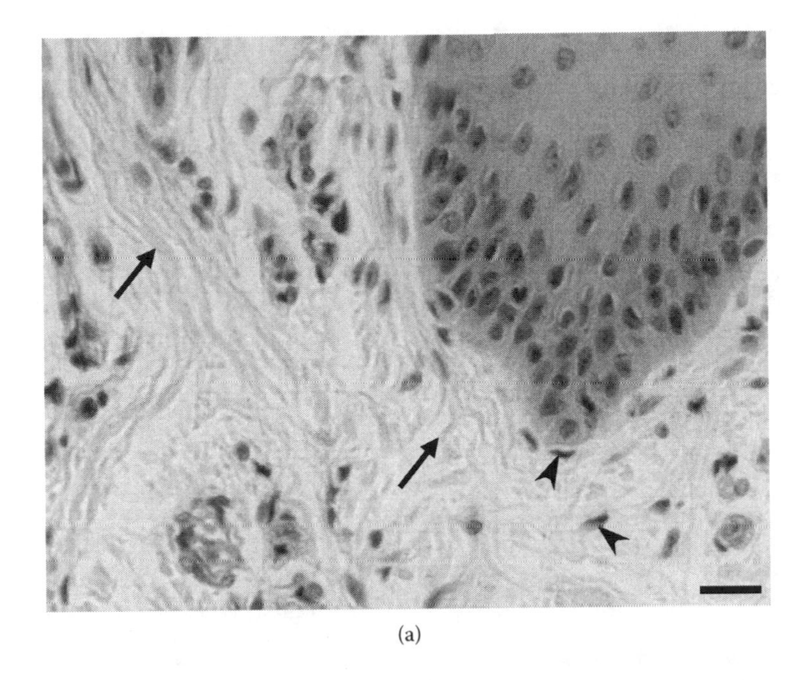

(a)

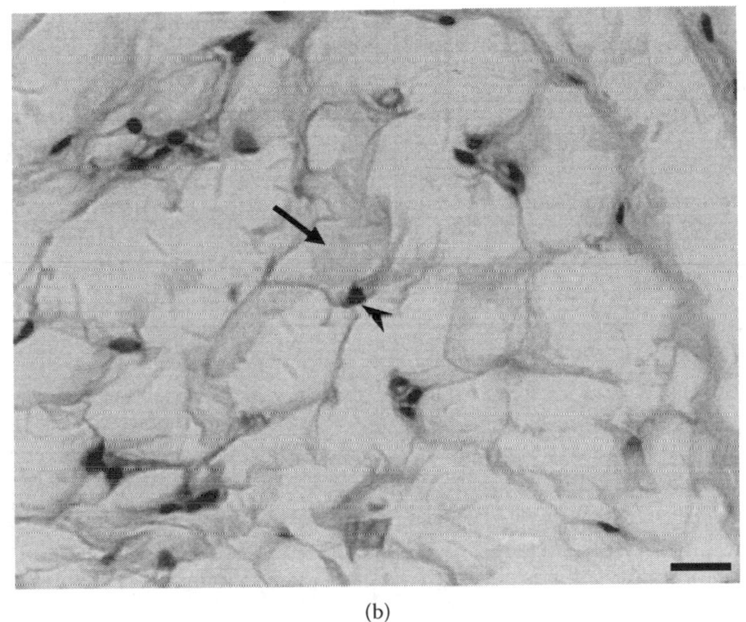

(b)

FIGURE 6.13
The ECM of the dermis and cells that dwell in it. (a) Collagen fibers and bundles (black arrows) form the matrix of the dermis and are particularly dense in the upper portion of the dermis. (Part of the epidermis is seen as a dark patch in upper right corner.) Fibroblasts (black arrow heads) are distributed throughout the collagen matrix. Magnification: 400. Scale bar: 20 µm. (b) The cell bodies of adipocytes (black arrow), found in the hypodermis, are extremely large, and their nuclei (black arrow head) tend to locate at the corners. Magnification: 400. Scale bar: 20 µm. All images are obtained using an inverted optical microscope set to brightfield.

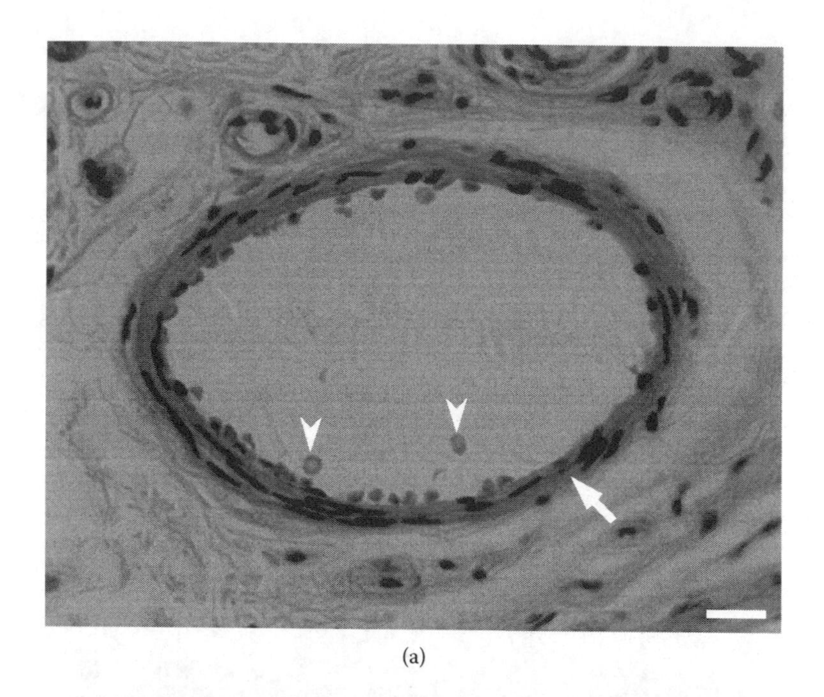

(a)

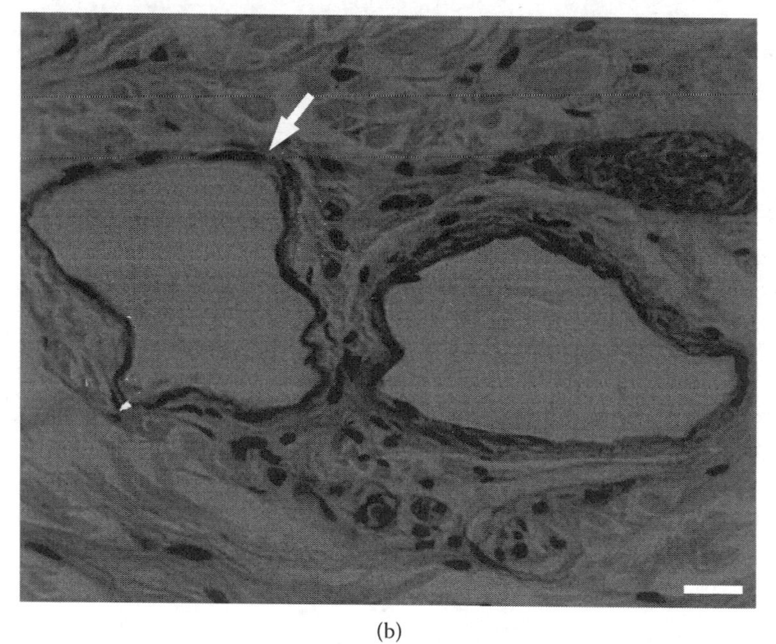

(b)

FIGURE 6.14
Blood vessels in the dermis. (a) An artery (white arrow) appears as a thick wall of smooth muscle cells with a lumen, in which some disc-shaped red blood cells remain (white arrow heads). (b) A vein (white arrow) has a thinner wall compared to an artery. For both images, magnification: 400. Scale bar: 20 μm. All images are obtained using an inverted optical microscope set to brightfield.

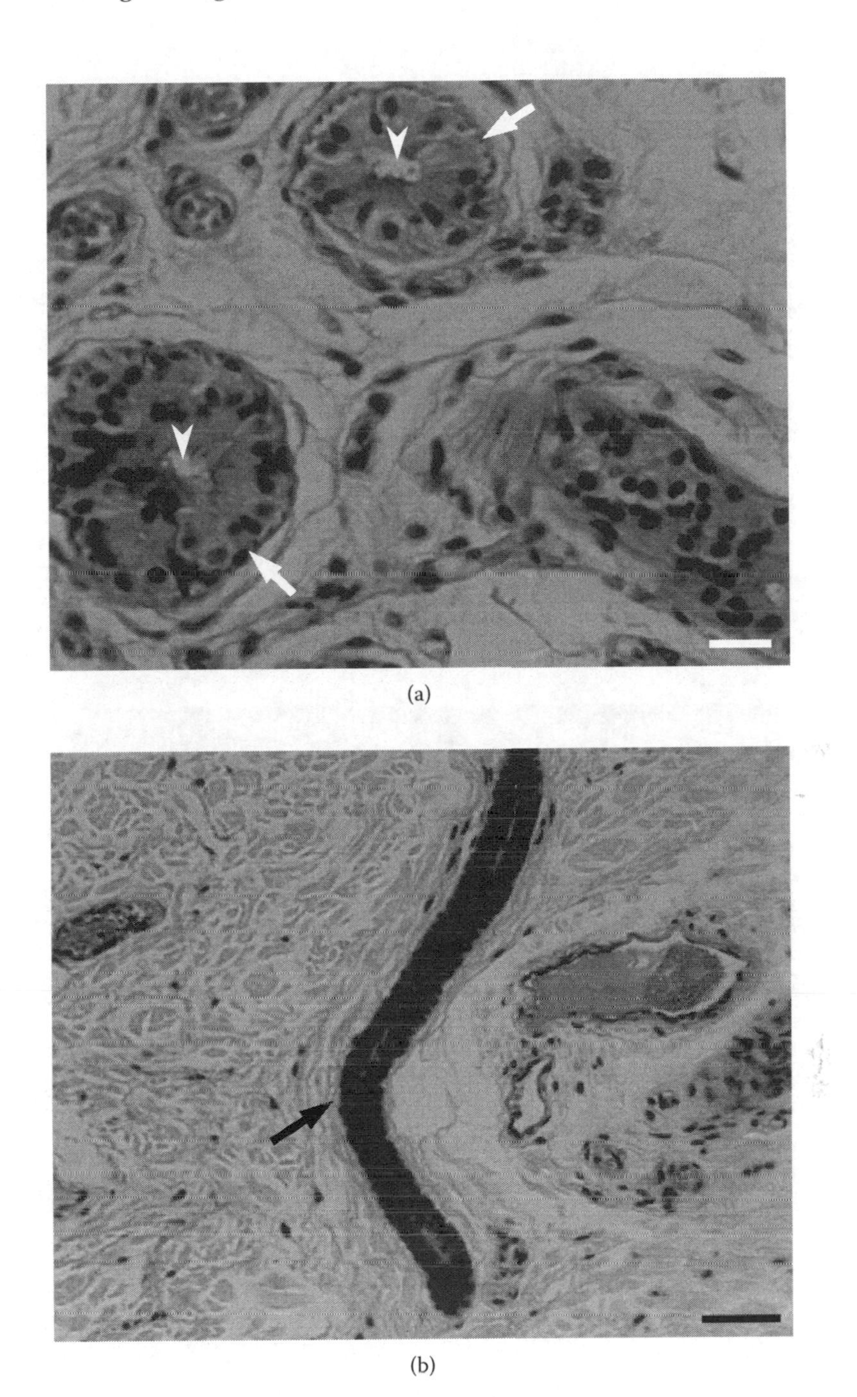

FIGURE 6.15
Eccrine sweat glands and duct in the dermis. (a) Sweat glands appear as clusters of darkly stained cells in radial-symmetrical, "chrysanthemum" style arrangement (white arrows); a lumen (white arrow head) is often visible in the middle of a gland. Magnification: 400. Scale bar: 20 μm. (b) A sweat duct (black arrow) embedded in the collagen matrix of the dermis. Magnification: 200. Scale bar: 50 μm. All images are obtained using an inverted optical microscope set to brightfield.

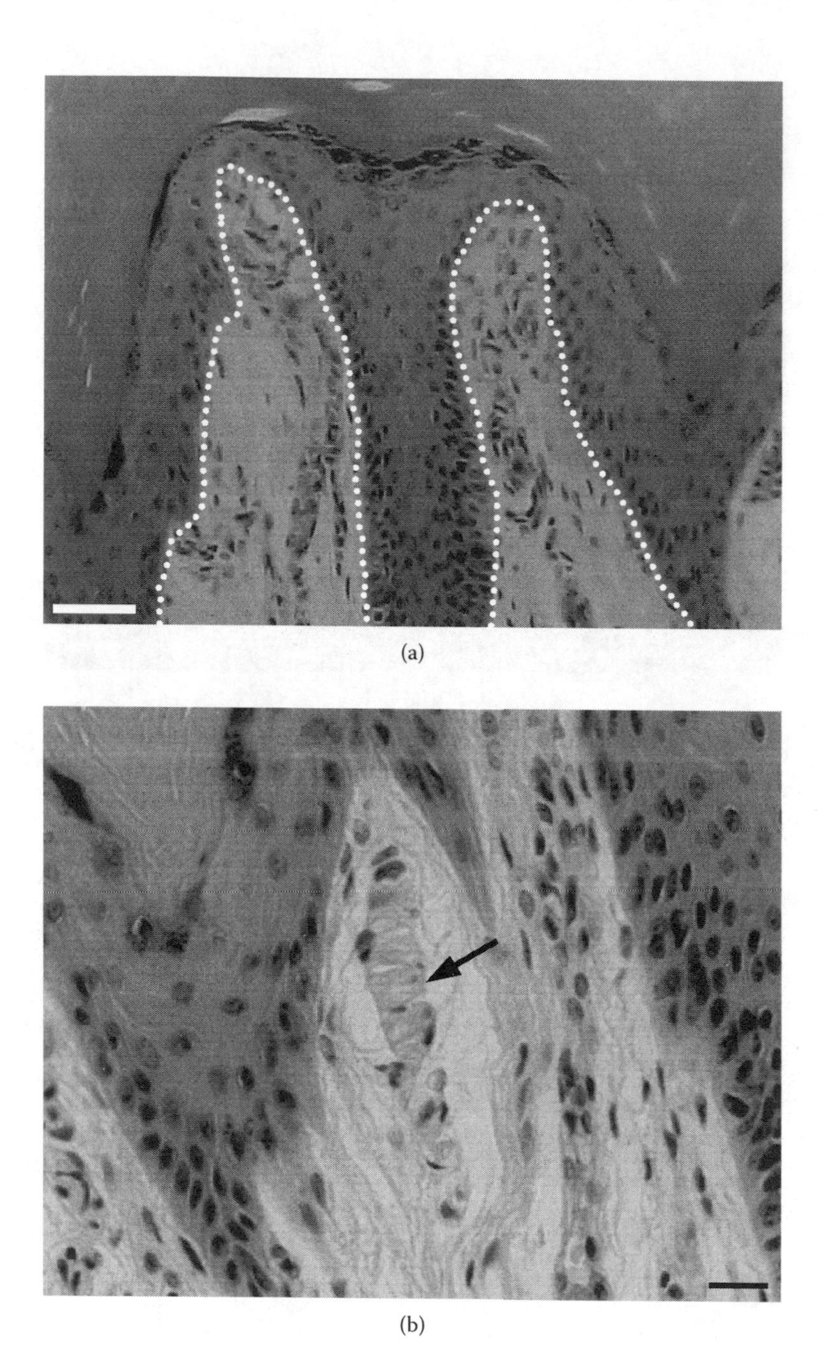

FIGURE 6.16
Dermal papillae and Meissner's corpuscle. (a) Dermal papillae (light colored) are outlined in white dotted line. Magnification: 200. Scale bar: 50 μm. (b) Meissner's corpuscle (black arrow) inside a dermal papilla. Magnification: 400. Scale bar: 20 μm. All images are obtained using an inverted optical microscope set to brightfield.

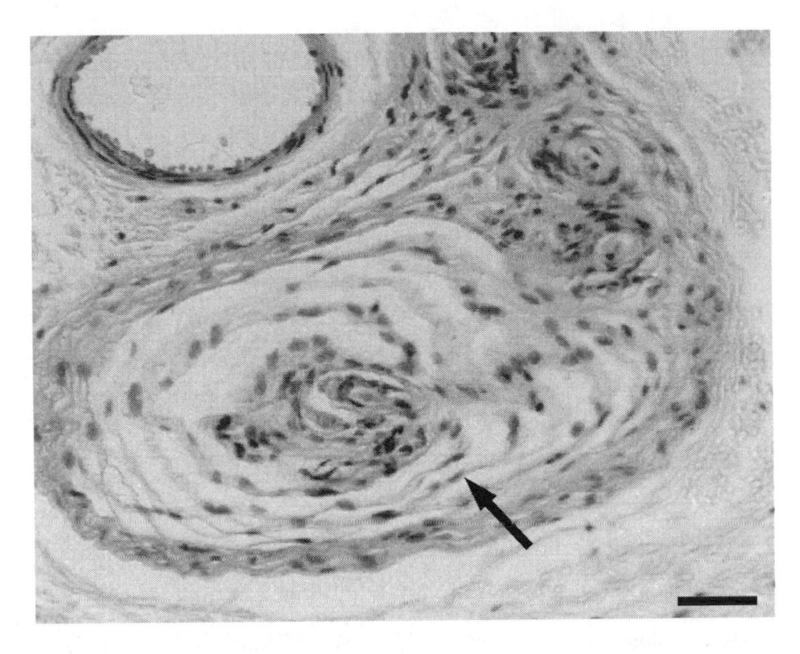

FIGURE 6.17
Pacinian corpuscle (black arrow), a nerve receptor that is characterized by the concentric arrangement of cells that gives it the look of a cross-sectioned onion. Magnification: 200. Scale bar: 50 μm. Image is obtained using an inverted optical microscope set to brightfield.

Part 2 (Day 1). Histology of Organotypic Skin Equivalent

1. *Samples and materials:* Histology slides of skin equivalents +TGF-α and −TGF-α, with hematoxylin and eosin (H&E) staining.
2. *The epidermis equivalent:* Identify the epidermis equivalent; note the differences between the shape of the cells near the basal lamina and that of the cells at the top layer. The epidermal layer, which consists of keratinocytes, is darkly stained (Figure 6.18). Take representative pictures of the epidermis.
3. *The dermis equivalent:* Identify the collagen matrix and the fibroblasts in the dermis equivalent; note the distribution of the cells and the collagen fibers (Figure 6.18).

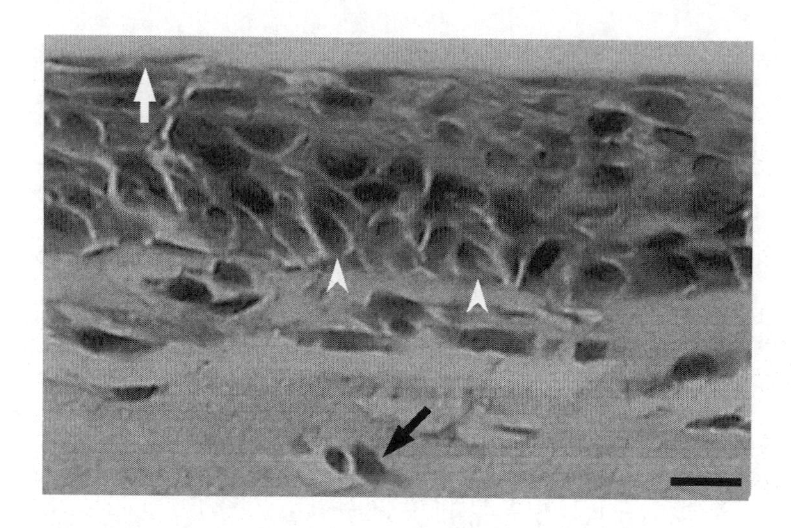

FIGURE 6.18
Histology of a tissue-engineered organotypic skin equivalent. The histology slide is prepared with cross-sections of the skin equivalent and stained with hematoxylin and eosin. The darkly stained upper portion is the epidermis equivalent, which is characterized by columnar basal cells (white arrow heads) and a stratified upper layer (white arrow). (The stratification is not yet complete in this case.) In the dermis equivalent, the collagen matrix is lightly stained; it is populated with fibroblasts that appear as dark spots (black arrow). Magnification: 400. Scale bar: 10 μm. Image is obtained using an inverted optical microscope set to brightfield.

Questions

1. In Session 1, why did we mix in 10X HBSS (with phenol red) into the collagen solution before titrating it with NaOH?
2. When creating the dermis equivalent, you and your co-workers neutralized the collagen solution with NaOH and added the fibroblast cells. After 1 hour of incubation, your group's collagen did not gel, but other groups' collagen had gelled and they used the same solution. Your instructor asked you if you saw white precipitation during the NaOH titration, and you did remember seeing some. Why was the white precipitation a problem?
3. You were titrating the collagen solution, and at some point you added a big drop of 2.0 *M* NaOH, and saw an instant change of color in the solution. Reassured by the color change as a sign of neutralization, you proceeded to add the fibroblast cells. The collagen gelled after incubation at 37°C. However, after 1 week, there did not seem to be any cell growth in the collagen matrix. What could have gone wrong?
4. Again, you were titrating the collagen solution with NaOH. This time you paid attention to the color and found that you had overshot it, so you back-titrated it with HCl. However, you overshot with the HCl and the solution turned yellow again, so you titrated it with NaOH again. After several back-and forth-titrations, you finally had a collagen solution with the right color—orange-red. (Turned out it wasn't all your fault; someone had made the concentrations of the NaOH and HCl solutions way too high.) You mixed in the fibroblast cells and incubated the mixture, and the collagen gelled. However, there seemed to be no cell growth (again!) after a week. What could have gone wrong this time?
5. When encapsulating fibroblast cells in collagen gel, what would likely happen if you forgot to place a glass ring on the gel?

6. You want to study the proliferation of the fibroblast cells encapsulated in collagen gel, so you seed collagen gels with a known number of cells on Day 1, and you plan to measure the number of cells in the collagen gels on Day 7. How would you make the measurements?
7. We added two reagents, ascorbic acid and TGF-α, to the skin-equivalent cultures. What are their respective functions?
8. When fixing the skin-equivalent cultures for histology studies, the samples become stiff after soaking in the fixation solution for 10 minutes. Why is that? Can this treatment be used to improve the mechanical properties of tissue-engineered skin-replacement materials in clinical applications?
9. Why must the embedding cassettes be labeled with pencils and not ink pens or permanent markers?

Appendix A. Sterilizing Collagen and Preparing Collagen Solution

Type I collagen is used in this module as a matrix material for the dermis equivalent. Since the final products—the skin equivalents—will not to be used for implantation, strict requirements for the collagen in terms of source, purity, antigenicity, or even the concentration of the collagen solution are not necessary, but two conditions are absolutely required: 1) the collagen solution must be able to form gel after neutralization, and 2) it must be *sterile*. To undergo a solution-to-gel transition, a collagen solution must have a high percentage of acid-soluble (that is, not cross-linked) and native (that is, not denatured) collagen. The collagen that we purified in Module II fits these criteria. Sterilizing collagen, on the other hand, requires special approaches because conventional sterilizing methods such as γ-radiation and heat treatment will cause cross-linking or denaturation of collagen. Sterile, ready-to-use type I collagen can be purchased from commercial vendors.* An economical alternative is to sterilize the collagen purified from calf skin in Module II using the following procedures. (If the amount of collagen purified by the students is not enough, the instructor will need to purify additional amount of collagen.)

Procedures

1. Preparation
   a. Samples and materials
      - Acid-soluble collagen, freeze-dried or in solution

---

* The following products might to be suitable. (This listing does not imply endorsement by the author.)

- Cellagen Solution AC-5: contains 0.5% solution (5 mg/ml) of acid-solubilized type I collagen in pH 3.0 solution, from MP Biomedicals (Solon, OH, www.mpbio.com), cat. no. 152394.
- BD Collagen I HC: contains rat-tail collagen in 0.02 *N* acid solution with concentration range 8–11 mg/ml, from BD Biosciences (San Jose, CA, www.bdbiosciences.com), cat. no. 354249.
- Calf-skin collagen in 0.075 *M* acetic acid with 0.01% thimerosal as a preservative with pH 3.7; concentration is approximately 10 mg/ml, from USB Corporation (Cleveland, Ohio, www.usbweb.com), cat. no. 13813. Note that this product contains thimerosal, which probably needs to be removed.

b. Reagents
    - 0.1 *M* acetic acid (prepare at least 300 ml)
    - De-ionized water
c. Special equipment and supplies
    - Regenerated cellulose dialysis tubing, MWCO 8000 Da., vol./cm ratio ~3 ml/cm
    - Two sets of dialysis tubing clamp closures
    - 2000-ml beaker
    - Heavy-duty aluminum foil
    - 1000-ml graduated cylinder
    - Scissors
    - Tweezers
    - Sterilization pouches
    - Millipore Steriflip™ (Millipore Corp.) sterile disposable vacuum filter units (Fisher cat. no. SCGP 005 25)
    - Bottle-top sterile filter unit
    - 500-ml media bottle
    - Stirring bar, ≥5 cm
    - Bell jar vacuum chamber

2. *Note:* Throughout the follow steps, keep the temperature ≤20°C if possible. At no time should the temperature be higher than 25°C. Also, keep the collagen solution and solid away from sunlight, UV light, or other strong light.
3. Weigh 1 g of freeze-dried collagen. (If the collagen is already in solution, proceed to Step 5.)
4. Dissolve the collagen in 50 ml of 0.1 *M* acetic acid. This may take overnight.
5. Distribute the collagen solution in two pieces of dialysis tubing, and dialyze it against de-ionized water in a 2000-ml beaker three times or until the collagen solution turns into solid gel.
6. Sterilizing supplies by autoclave: Place a stir bar inside the 1000-ml graduated cylinder and cover it with double-layered heavy-duty aluminum foil. Autoclave the graduated cylinder with the aluminum foil cap, a pair of scissors, and a pair of tweezers sealed in sterilization pouches, and a 500-ml media bottle with a stirring bar inside.
7. Inside a BSC, sterile-filter de-ionized water with bottle-top sterile filter unit, and then add 750 ml ethanol and 250 ml sterile-filtered water to the sterilized graduated cylinder (From Step 6). (Save the aluminum foil cap and keep it sterile.)
8. Take the dialyzed collagen out of the beaker (From Step 5), blot off the excess water, and spray the outside of the tubing and the clamps with 70% ethanol all over. After this aseptic treatment, move the two tubes of collagen into the BSC and place them into the 75% ethanol solution in the graduated cylinder (From Step 7).

Seal the top of the graduated cylinder with the aluminum foil cap. (Note that the seal will not be airtight, but it should be pressed down as tightly as possible.)

9. Dialyze the collagen in 75% ethanol with stirring for 3 days to sterilize.
10. Aseptically move a pair of sterilized scissors and two Steriflip sterile filter units inside the BSC. Move the two tubes of collagen in dialysis tubing into the BSC as well. Set up a regular sterile 50-ml tube in a stand, and then cut open one end of a tube of collagen and "dump" the collagen gel into the 50-ml tube. Top the tube with ethanol solution from the 1000-ml cylinder, and screw the Steriflip filter unit on top. Attach the filter to vacuum, flip the two tubes upside down, and turn on the vacuum. The alcohol should be filtered into the Steriflip 50-ml tube. Afterwards, unscrew the filter along with the regular 50-ml tube (which contains the collagen). Reverse the filter on the regular tube by unscrewing it, flipping it upside down, and screwing it back onto the tube (Figure 6.19). The collagen should remain sterile as long as the tube is capped with the filter. Repeat these procedures with the other tube of collagen.
11. Place the tubes with collagen in a bell jar vacuum chamber that is connected to a vacuum or water aspirator pump. Leave the aspirator on overnight to remove alcohol from the collagen gel.
12. Place the collagen in a freezer-dryer and let it dry overnight. (It is not necessary to freeze the collagen.) The collagen gels will likely shrink into small pellets after drying.
13. Aseptically move the autoclaved 500-ml media bottle inside a BSC. Attach a bottle-top sterile vacuum filter unit, and filter 200 ml of 0.1 *M* acetic acid into the bottle.
14. Aseptically move the sterilized and dried collagen inside the BSC. With a pair of sterilized tweezers, add the collagen pellets into the 500-ml media bottle. (If you need to know the precise weight of the collagen at this point, then weigh a sterile 50-ml tube, transfer the collagen pellets into the tube, and weigh the tube again.)
15. Stir the solution in the 500-ml media bottle with medium speed until the collagen pellets are completely dissolved. This may take overnight.
16. The collagen solution should have a final concentration of ~5 mg/ml. It is good for years when stored at 4°C and keep it sterile. Do not freeze the solution.

Appendix B. Custom-Made Teflon Support for Air Exposure of the Epidermis Equivalent

Air exposure is the key to stratification of the epidermis equivalent. In practice, air exposure is achieved by lowering the culture medium level to below the epidermis equivalent. If tissue culture inserts are used with a 6-well plate for the skin-equivalent culture, inconvenience arises when the medium level is reduced in the plate well for the air exposure. The volume of culture medium in the well would become too low to maintain the cells for the conventional medium change interval of 2–3 days; medium change needs to be performed

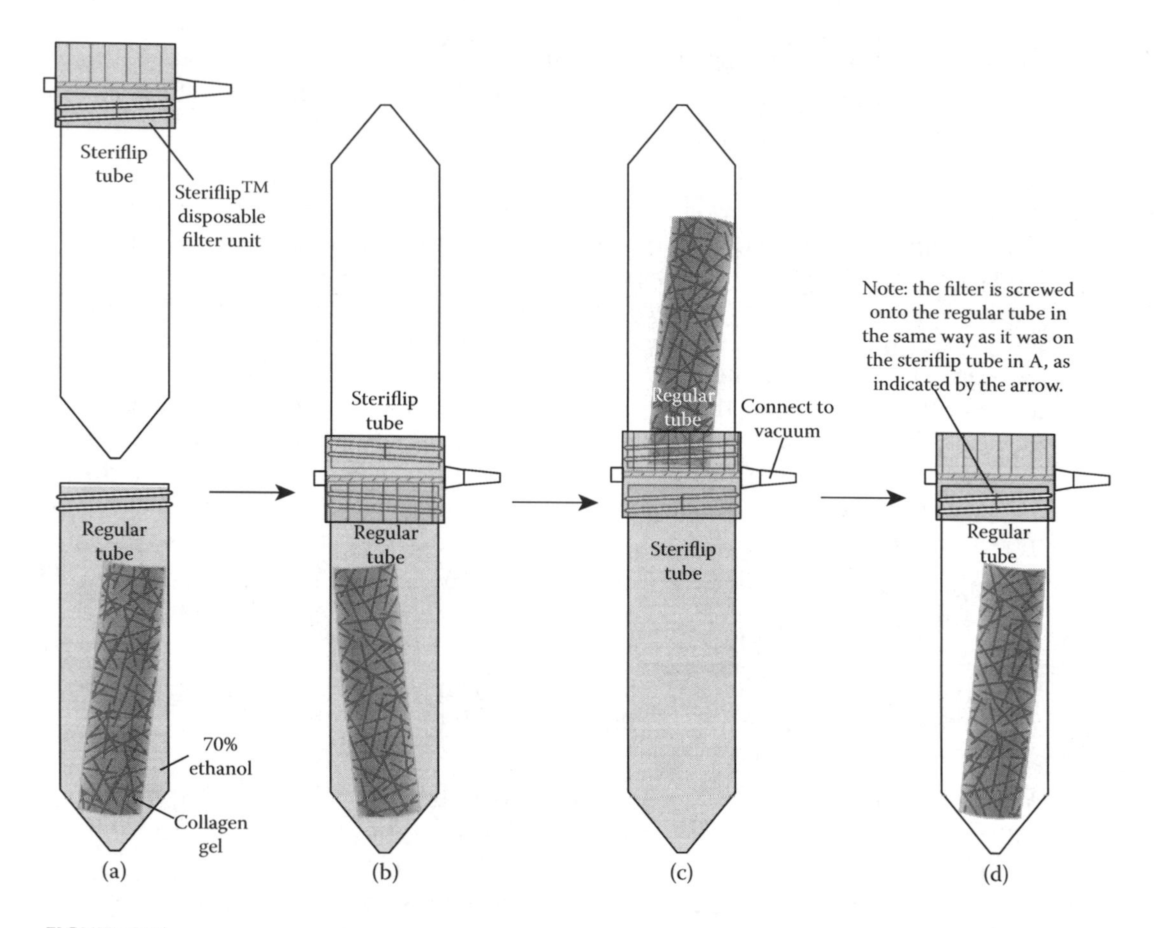

FIGURE 6.19
Sterilizing collagen using a Millipore Steriflip disposable filter unit. The collagen has been dissolved in acidic solution, dialyzed in de-ionized H_2O, and then dialyzed in 75% ethanol up to this point. (a) Transfer the collagen to a sterile 50-ml tube under aseptic conditions and fill it with 70% ethanol; unwrap a Steriflip filter unit. (Note that an arrow is drawn on the filter to indicate its orientation.) (b) Screw the Steriflip filter unit on top of the tube that contains the collagen. (c) Flip the two tubes upside down and then connect the filter to vacuum. The alcohol solution should be filtered to the Steriflip tube. When finished, unscrew the filter but do not detach the collagen-containing tube. (d) Finally, flip the filter to cap the collagen-containing tube. The collagen can then be aspirated to remove the alcohol and dried in a vacuum to remove residual water while kept sterile inside the filter-capped tube.

every day to supply enough nutrients and remove wastes. With the inconvenience also comes the increased risk of contamination.

There are several different solutions to this problem. For example, a specially designed bioreactor with a membrane-bottomed chamber suspended in a well is used for the industrial production of organotypic skin replacements (Roos et al. 2004). For student experiments, custom-made, reusable Teflon supports based on the idea of "suspended chamber" can be an economic solution (Figure 6.20). These supports are designed to be used with tissue culture inserts for a 6-well plate: Skin equivalents are cultured in inserts placed in the 6-well plate up to the point of air exposure for the epidermis equivalent, and then the inserts are placed in the Teflon supports and cultured in containers that allow large volume of culture medium and require less frequent medium changes compared to the 6-well plate (Figure 6.21).

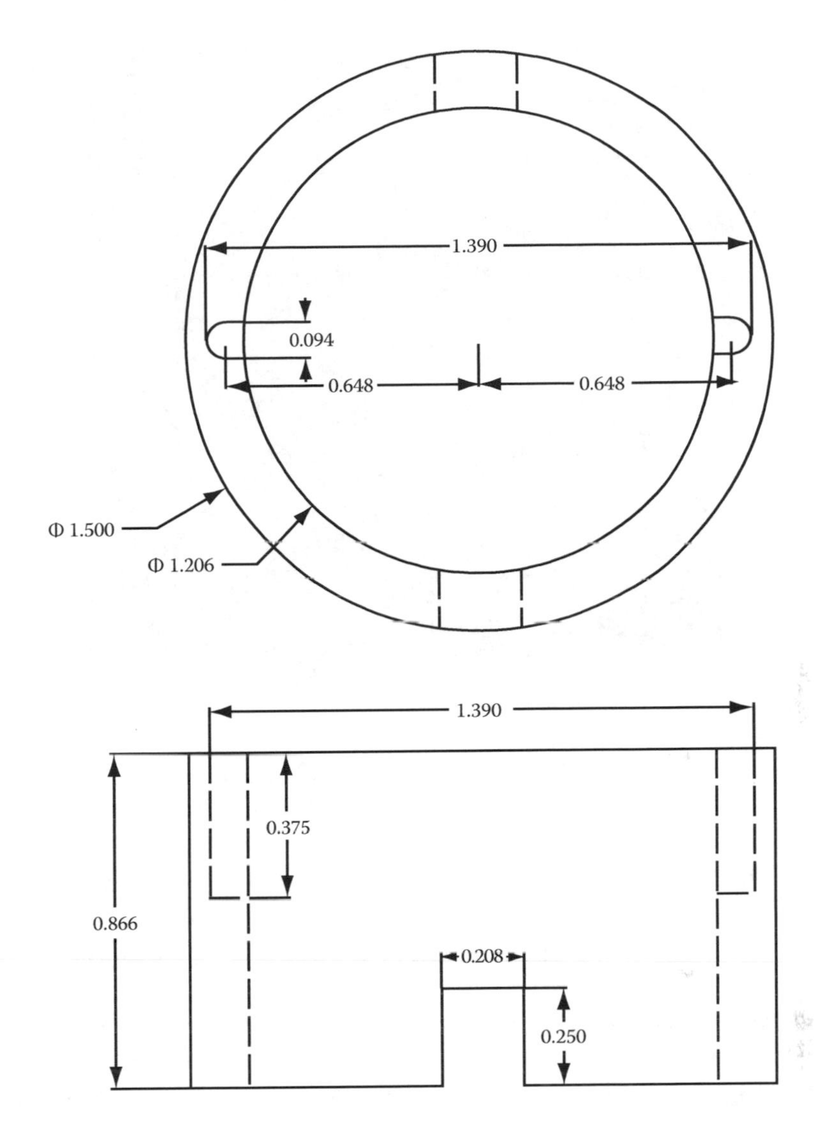

FIGURE 6.20
Machine shop drawings of custom-made Teflon supports for use with BD Falcon cell culture inserts for 6-well plates (cat. no. 353091). The supports can be machined from 1.5″ diameter PTFE rod (McMaster Carr, cat. no. 2905K16).

Appendix C. Recipes and Sources for Equipment, Reagents, and Supplies*

Session 1. The Dermis Equivalent

- **Collagen.** See Appendix A.
- **NIH 3T3 cells.** American Type Culture Collection (Manassas, VA), ATCC no. CRL-1658.

* *Disclaimer*: Commercial sources for reagents listed are used as examples only. The listing does not represent endorsement by the author. Similar or comparable reagents can be purchased from other commercial sources. See selection criteria.

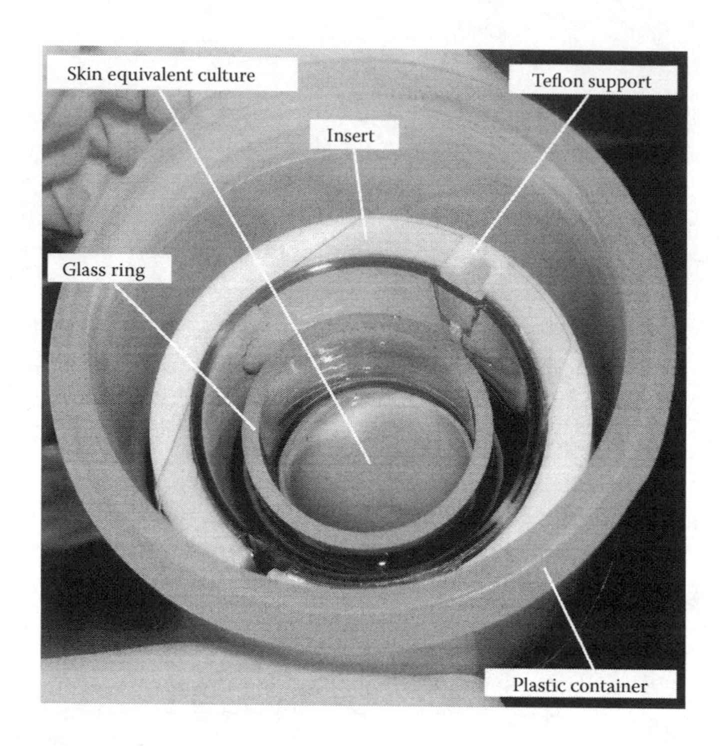

FIGURE 6.21
An organotypic skin-equivalent culture in a tissue culture insert with a custom-made Teflon support in a polypropylene container. A glass ring is seen on top of the culture. The epidermis equivalent is to be air exposed after removal of the glass ring.

- **0.1 *M* acetic acid solution.** Measure 350 ml of de-ionized water with a graduated cylinder; add 2.0 ml of glacial acetic acid.
- **10X HBSS with phenol red.** Sigma-Aldrich (St. Louis, MO), cat. no. H4641. *Sel. crit.:* Contains phenol red.
- **DMEM.** Thermo Scientific Hy Clone™ DMEM, with 4g/L glucose, 2 mM L-glutamine and sodium pyruvate. Fisher cat. no. SH30243FS. *Sel. crit.:* Similar compositions.
- **Complete DMEM (1 liter).** 890 ml DMEM, 100 ml FBS, and 10 ml 100X penicillin-streptomycin stock solution.
- **2.0 *M* NaOH.** Add 20 ml of 10 *M* sodium hydroxide solution (Sigma-Aldrich, cat. no. 72068) to 80 ml of sterile-filtered de-ionized water. ***Caution:*** NaOH solutions are extremely caustic. *Sel. crit.:* The concentration does not need to be strictly accurate.
- **1.0 *M* HCl.** Sigma-Aldrich cat. no. H9892. *Sel. crit.:* The concentration does not need to be strictly accurate.
- **L-Ascorbic acid stock solution (5 mg/ml), 100X.** Weigh 500 mg of L-ascorbic acid powder (Sigma-Aldrich, cat. no. A4544) and dissolve in 100 ml de-ionized water. Sterile filter with bottle-top or syringe-tip filter. Store in 10-ml aliquots at –20°C. Keep away from light.
- **Complete DMEM supplemented with 50 mg/ml ascorbic acid (100 ml).** Add 1.0 ml of 5 mg/ml L-ascorbic acid stock solution to 100 ml of complete DMEM.
- **Tissue culture insert.** BD Falcon, BD Biosciences, mfr. no. 353091, Fisher cat. no. 08-771-3. *Sel. crit.:* For 6-well plates, membrane pore size >1 μm.

- **Glass ring.** Custom-made in local glass shop, o.d. 22 mm, thickness 1.5 mm, height 15.5 mm. *Sel. crit.:* Similar dimensions.

Session 2. The Epidermis Equivalent

- **PBS with 0.05% EDTA.** Add 250 mg of EDTA (Sigma-Aldrich, cat. no. E5134) to 500 ml of PBS. Sterile filter with bottle-top 0.22 μm filter unit.
- **TGF-α stock solution (10 μg/ml), 5000X.** Dissolve 100 μg of recombinant human TGF-α (BioVision, Mountain View, CA, cat. no. 4339-100) in 10 ml PBS. Sterile filter with syringe-tip 0.22 μm filter. Store as 1-ml aliquots at –20°C. To minimize freezing and thawing, if a 1-ml aliquot is thawed but not all of it is used within 1 week, store the rest of it in 100-μl aliquots at –20°C right away.
- **Complete DMEM with 2 ng/ml TGF-α.** Add 20 μl of 10 μg/ml TGF-α stock solution to 100 ml of complete DMEM.
- **Teflon support.** Custom-made in local machine shop, dimensions in Figure 6.20.
- **Container (for epidermis air-exposure).** Nalegene straight-side polypropylene jar with screw cap, 2 oz., mfr. no. 2118-0002, Fisher cat. no. 11-815-10B. *Sel. crit.:* Similar dimensions, autoclavable.
- **Fixation solution (3.7% formaldehyde in PBS).** In a fume hood, add 50 ml of 37% formaldehyde solution (Sigma-Aldrich, cat. no. 533998) to 450 ml of PBS.

Session 3. Histological Studies of Natural Skin and Organotypic Skin Equivalent

- **Human skin slide.** Home Science Tools (Billings, MT), product no. MS-HUSKIN. *Sel. crit.:* From human thick skin.

References

Bello, Y.M., and A.F. Falabella, The role of graftskin (Apligraf®) in difficult-to-heal venous leg ulcers, *J. Wound Care*, 11, 182–183, 2003.

Bouchie, A., Tissue engineering firms go under, *Nat. Biotechnol.*, 20, 1178–1179, 2002.

Boyce, S., R. Kagan, D. Greenhalgh, P. Warner, K. Yakuboff, T. Palmieri, and G. Warden, Cultured skin substitutes reduce requirements for harvesting of skin autograft for closure of excised, full-thickness burns, *J. Trauma.*, 60, 821–829, 2006.

Clark, R.A.F., K. Ghosh, and M.G. Tonnesen, Tissue engineering for cutaneous wounds, *J. Invest. Dermatol.*, 127, 1018–1029, 2007.

Griffith, M., R. Osborne, R. Munger, X. Xiong, C.J. Doillon, N.L.C. Laycock, M. Hakim, Y. Song, and M.A. Watsky, Functional human corneal equivalents constructed from cell lines, *Science*, 286, 2169–2172, 1999.

Haddow, D., D. Steele, R. Short, R. Dawson, and S. Macneil, Plasma-polymerized surfaces for culture of human keratinocytes and transfer of cells to an *in vitro* wound-bed model, *J. Biomed. Mater. Res. A*, 64, 80–87, 2003.

Hermans, M.H., Clinical experience with glycerol-preserved donor skin treatment in partial thickness burns, *Burns Incl. Therm. Inj.*, 15, 57–59, 1989.

Hu, J.C., and K.A. Athanasiou, A self-assembling process in articular cartilage tissue engineering, *Tissue Eng.*, 12, 969–979, 2006.

Jarman-Smith, M., T. Bodamyali, C. Stevens, J. Howell, M. Horrocks, and J. Chaudhuri, Porcine collagen crosslinking, degradation and its capability for fibroblast adhesion and proliferation, *J. Mater. Sci. Mater. Med.*, 15, 925–932, 2004.

Kumar, R.J., R.M. Kimble, R. Boots, and S.P. Pegg, Treatment of partial-thickness burns: A prospective randomized trial using Transcyte™, *ANZ J. Surg.*, 74, 622–626, 2004.

Limat, A., L.E. French, L. Blal, J.H. Saurat, T. Hunziker, and D. Salomon, Organotypic cultures of autologous hair follicle keratinocytes for the treatment of recurrent leg ulcers, *J. Am. Acad. Dermatol.*, 48, 207–214, 2003.

Lipkin, S., E. Chaikof, Z. Isseroff, and P. Silverstein, Effectiveness of bilayered cellular matrix in healing of neuropathic diabetic foot ulcers: Results of a multicenter pilot trial, *Wounds*, 15, 230–236, 2003.

Maas-Szabowski, N., A. Stärker, and N.E. Fusenig, Epidermal tissue regeneration and stromal interaction in HaCaT cells is initiated by TGF-α, *J. Cell. Sci.*, 116, 2937–2948, 2003.

MacNeil, S., Progress and opportunities for tissue-engineered skin, *Nature*, 445, 874–880, 2007.

Marston, W.A., J. Hanft, P. Norwood, and R. Pollak, The efficacy and safety of Dermagraft in improving the healing of chronic diabetic foot ulcers: Results of a prospective randomized trial, *Diabetes Care*, 26, 1701–1705, 2003.

Navarro, F., M. Stoner, C. Park, J. Huertas, H. Lee, F. Wood, and D. Orgill, Sprayed keratinocyte suspensions accelerate epidermal coverage in a porcine microwound model, *J. Burn Care Rehabil.*, 21, 513–518, 2000.

NIH (National Institutes of Health), PAR-06-504: Enabling technologies for tissue engineering and regenerative medicine, Department of Health and Human Services, Washington, DC, 2006.

O'Connor, N.E., J.B. Mulliken, S. Banks-Schlegel, O. Kehinde, and H. Green, Grafting of burns with cultured epithelium prepared from autologous epidermal cells, *Lancet*, 1(8211), 75–78, 1981.

Pastore, S., F. Mascia, V. Mariani, and G. Girolomoni, The epidermal growth factor receptor system in skin repair and inflammation, *J. Invest. Dermatol.*, 128, 1365–1374, 2007.

Roos, E., C. O'Reilly, R. Chevere, and L.M. Wilkins, U.S. patent no. 6,730,510 B2: Culture dish and bioreactor system, Organogenesis, Inc., Canton, MA, 2004.

Russell, S.B., J.D. Russell, and K.M. Trupin, Collagen synthesis in human fibroblasts: effects of ascorbic acid and regulation by hydrocortisone, *J. Cell. Physio.*, 109, 121–131, 1981.

Shen, G., H.C. Tsung, C.F. Wu, X.Y. Liu, X.Y. Wang, W. Liu, L. Cui, and Y.L. Cao, Tissue engineering of blood vessels with endothelial cells differentiated from mouse embryonic stem cells, *Cell Res.*, 13, 335–341, 2003.

Stark, H.-J., A. Szabowski, N.E. Fusenig, and N. Maas-Szabowski, Organotypic cocultures as skin equivalents: A complex and sophisticated *in vitro* system, *Biol. Proced. Online*, 6, 55–60, 2004.

Stern, R., M. McPherson, and M.T. Longaker, Histologic study of artificial skin used in the treatment of full thickness thermal injury, *J. Burn Care Rehabil.*, 11, 7–13, 1990.

Tausche, A., M. Skaria, L. Böhlen, K. Liebold, J. Hafner, H. Friedlein, M. Meurer, R. Goedkoop, U. Wollina, D. Salomon, and T. Hunziker, An autologous epidermal equivalent tissue-engineered from follicular outer root sheath keratincoytes is as effective as split-thickness skin autograft in recalcitrant vascular leg ulcers, *Wound Repair Regen.*, 11, 248–252, 2003.

Wainwright, D., M. Madden, A. Luterman, J. Hunt, W. Monafo, D. Heimbach, R. Kagan, K. Sittig, A. Dimick, and D. Herndon, Clinical evaluation of an acellular allograft dermal matrix in full- thickness burns, *J. Burn Care Rehabil.*, 17, 124–136, 1996.

Wright, K., K. Nadire, P. Busto, R. Tubo, J. McPherson, and B. Wentworth, Alternative delivery of keratinocytes using a polyurethane membrane and the implications for its use in the treatment of full-thickness burn injury, *Burns*, 24, 7–17, 1998.

7

Module V. Bioceramics: Porous Hydroxyapatite Composite

Bioceramics are ceramics materials that are used for medical and dental applications. They are categorized according to their bioactivities as inert, resorbable, or surface active (Figure 7.1) (Vallet-Regí 2001). Alumina is an example of inert bioceramic: It is a strong material that has been used in the balls and cups of hip prostheses and is noted for its wear resistance; it is inert, therefore implants made with alumina do not bond with surrounding tissues. On the other hand, alumina is nonbiodegradable, thus it is not suitable for resorbable implants. Bioactive glasses are surface active bioceramics, and they are also largely nonresorbable; they are amorphous, strong, and bioactive, and allow bonding between implants and tissues and sometimes ingrowth of tissues in the implants. Resorbable bioceramics are largely based on calcium and phosphate. These materials are versatile, biocompatible, mostly resorbable, and very bioactive, although they lack the mechanical strength of the inert bioceramics or the bioactive glasses. Currently calcium phosphate-based bioceramics are used for applications such as implant coating, bone repair and void-filling, dental repair, and drug delivery.

One of the most important calcium phosphates is hydroxyapatite (HA), the main mineral content of bones and teeth. It has long been used for coating titanium hip-joint stems for better integration between the implant and the femur bone, as well as numerous applications as bone void fillers. The chemical formula of HA is usually expressed as $Ca_{10}(PO_4)_6(OH)_2$, denoting the number of Ca, P, O, and H atoms that make up a crystallographic *unit cell* (Figure 7.2) (de Leeuw 2001). Notice that in this unit cell, the four positions

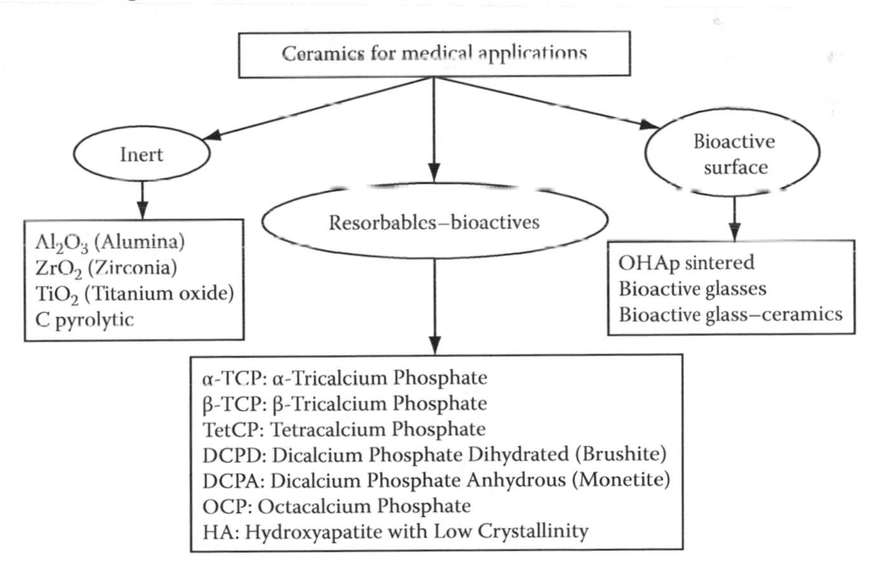

FIGURE 7.1
Types of bioceramics as defined by activity. Reproduced by permission of the Royal Society of Chemistry (Vallet-Regí 2001).

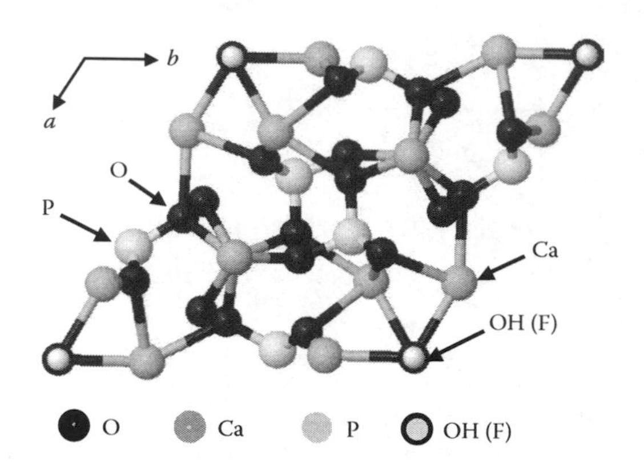

FIGURE 7.2
Structure of the hexagonal apatite unit cell. The *c* direction is perpendicular to the plane of the paper. The positions occupied by hydroxy ions in this unit cell can be substituted by fluoride, chloride, or carbonate ions. Reproduced by permission of the Royal Society of Chemistry (de Leeuw 2001).

occupied by OH can be substituted by other anions such as F^-, Cl^-, and CO_3^{2-}, and indeed, the mineral phase of bones and teeth consists of a mixture of HA, its substituted variants, and other forms of calcium phosphates. For this reason, even though the Ca:P ratio for pure HA is 1.67, in bone mineral the Ca:P ratio ranges from 1.37 to 1.87. HA and its substituted variants are extremely stable chemically in neutral or basic environments, but not in the presence of acid. This is the reason caries form in teeth as a result of HA erosion, which occurs when bacteria infection creates an acidic environment on the teeth. HA also has remarkable thermal stability: It can resist thermal decomposition at temperatures up to 1000°C, allowing sintering treatments that modify the crystallinity and porosity of the HA material; such treatments are routinely used to coat metal implants with HA. Currently, both natural and synthetic HA are available. Bones and corals are natural sources for HA, but for some applications, synthetic HA is preferred because the composition and crystallinity can be controlled. HA with lower crystallinity is found to be more bioactive.

As mentioned above, the structure of HA can be easily substituted with F^-, and the result is fluorapatite (FA). Compared to HA, FA is even less soluble, more resistant to chemical degradation, and has better thermal stability. The remarkable chemical stability of FA is the main reason fluorination is used to prevent dental caries: It works by inhibiting the demineralization and promoting the remineralization of the enamel layer of tooth. FA also stores F^-, which has been shown to stimulate proliferation of bone-forming cells (Farley et al. 1983); the potential of FA as a bone-replacement material is promising.

For orthopedic applications such as bone void filling, porous HA materials are often used in order to repair damage and promote new bone growth. These materials must be mechanically strong and they must be osteoconductive—supporting bone cell migration, attachment and proliferation and promoting vascularization, and the eventual new bone formation. While HA has excellent osteoconductivity, the mechanical strength of HA itself is relatively weak, and nonductile; therefore, composites of HA with polymer materials have been extensively studied to improve the mechanical properties of HA-based biomaterials.

In this module, we will study the chemical and mechanical properties of HA and its potential as a bone-replacement material. We will also critically compare HA with FA. We will begin our experiments with synthesis of HA and/or FA using a wet precipitation method. The synthesized HA and FA will be used to fabricate porous HA/FA-polymer composites; and the mechanical properties and acid degradation of these materials will be studied. Our objective is to gain a better understanding of the concept of bioceramics, the important properties of bioceramic materials, and how these properties are used in biomedical applications.

Session 1. Synthesis of HA and FA

Among the many methods for HA synthesis, wet precipitation is the simplest and probably the most cost effective. In this method, a calcium salt and a phosphate salt are both dissolved or suspended before mixing at a stoichiometric Ca:P ratio of 1.67, the same as that of HA, and the reaction is allowed to proceed at a basic pH until HA formation is complete. Since factors such as the type of precursor salts, their respective concentrations in solution or suspension, the reaction temperature, and the reaction pH will all likely affect the quality of the final product, the reaction conditions need to be carefully monitored.

In this session, we will use calcium hydroxide and (ortho-)phosphoric acid as precursors to synthesize HA according to the following reaction:

$$10Ca(OH)_2 + 6H_3PO_4 \rightarrow Ca_{10}(PO_4)_6(OH)_2 \downarrow +18H_2O$$

In this reaction, the pH is a critical factor for the formation of HA. At pH 10 to 11, the formation of HA includes the following steps (Liu et al. 2001):

$$\text{OCP} \Rightarrow \text{ACP} \rightarrow \text{DCP} \rightarrow \bar{\text{HA}}$$

The abbreviations represent the following intermediates:

OCP: Otacalcium phosphate, $Ca_8H_2(PO_4)_6\,5H_2O$.
ACP: Amorphous calcium phosphate, $Ca_3(PO_4)_2{\cdot}xH_2O$.
DCP: Calcium-deficient hydroxyapatite, $Ca_{10-z}(HPO_4)_z(PO_4)_{6-z}(OH)_{2-z}{\cdot}nH_2O$, $0 \le z \le 1$.

The formation of OCP and the conversion of OCP to ACP happen very quickly (on a time scale of minutes), but the kinetics for the rest of the reaction is relatively slow (on a time scale of hours to tens of hours) and is determined largely by the reaction temperature. Higher temperature leads to faster reaction kinetics and yields HA with a higher level of crystallinity.

An advantage of the wet precipitation synthesis method is the easy substitution of HA with fluoride, the result of which is FA. Fluoride can be simply added to the reaction for HA to form FA according to the following equation:

$$10Ca(OH)_2 + 6H_3PO_4 + xNH_4F \rightarrow Ca_{10}(PO_4)_6(OH)_{2-x}F_x \downarrow +18H_2O + xNH_4OH,\ 0 \le x \le 1$$

The degree of fluorination can be controlled by the amount of the added fluoride. This allows the chemical and physical properties of the HA/FA product to be "tuned" for specific applications.

Procedures

Part 1 (Day 1). The Wet Precipitation Reaction

1. *Check-in*
   a. Reagents
      - Orthophosphoric acid, H_3PO_4 (MW 98.0), 0.60 *M*
      - Calcium hydroxide, $Ca(OH)_2$ (MW 74.09), solid
      - Ammonium fluoride, NH_4F (MW 37.04), solid
      - NaOH, 2.0 *M*
      - pH test papers
   b. Special equipment and supplies
      - 250-ml beaker
      - 150-ml flask
      - 100-ml cylinder
      - Spatulas, large and small
2. *Preparing phosphoric acid*
   a. *Note:* Phosphoric acid solution is usually prepared by diluting from 80–85% concentrated acid. The accurate concentration of the phosphoric acid should be determined before lab by your instructor.
   b. Place a stir bar in a 250-ml beaker, and add 80 ml of the 0.60 *M* phosphoric acid solution using a graduated cylinder.
3. *Preparing* $Ca(OH)_2$
   a. ***Caution:*** Calcium hydroxide is an irritant fine powder. Handle with care to minimize dust and avoid inhaling.
   b. Use the actual concentration of the 0.6 *M* phosphoric acid solution to calculate the weight of calcium hydroxide according to the stoichiometric Ca:P ratio of 1.67:

   $$W_{Ca(OH)2} = ([\text{phosphoric acid}] \times 0.080 \times 1.67) \times 74.09 \text{ (g)}.$$

   Weigh the calculated amount of calcium hydroxide powder.
4. *The reaction*
   a. Place the beaker with phosphoric acid on a stirring plate and start stirring at medium speed. Use a spatula to add a small amount of $Ca(OH)_2$ to the phosphoric acid solution; wait until the solid is mostly dissolved or dispersed, and then repeat until the calcium hydroxide is completely added to the reaction solution. (To synthesize FA, see Step 5.) Observe what happens in the beaker carefully.
   b. Place a strip of pH test paper on a small Kimwipe tissue. Dip a plastic transfer pipette into the solution (no need to squeeze the bulb), and then dab its tip on the pH test paper. Compare the color to the standard color chart (on the pH

paper container) to decide the pH. If the pH is <10, then titrate in 2.0 *M* NaOH drop by drop to bring the pH up to 10.

c. Monitor and adjust the pH according to Step 4b every 5 minutes for 20 minutes.

5. *Optional fluoride substitution*: ***Caution:*** Ammonia will be released when NH_4F is added to the reaction mixture. Add 0.60 g of NH_4F to the reaction after Step 4a. Check the pH according to Steps 4b and c. (*Note:* Half of the groups in the class can be assigned to synthesize HA and the other half FA.)
6. *Incubation:* Properly label the beaker and cover it with Saran wrap. Leave the reaction mixture stirring until the next lab session. Your instructors will check the pH of your group's reaction mixture daily to keep the pH >10 during the incubation. This incubation period allows HA (or FA) to form and mature. (The reaction product is "mature" in two aspects: chemically it means that the Ca:P ratio has reached a stable value, and physically the HA or FA crystallites have reached stable sizes.)

Part 2 (Day 2). Harvesting and Drying HA/FA

1. *Check-in*
   a. Samples and materials
      - Your group's reaction mixture from Part 1
   b. Reagents
      - 70% ethanol
      - 100% ethanol
   c. Special equipment and supplies
      - Two 50-ml conical centrifuge tubes
      - Centrifuge with swinging bucket rotor
      - A plastic stirring rod
      - Large (100 mm) ceramic Büchner funnel (handle with care!)
      - 1000-ml filtration flask
      - Water aspirator pump
      - Filter papers, diameter 90 mm
      - Rubber adaptor for the filtration flask
      - Large watch glass (Figure 7.3)

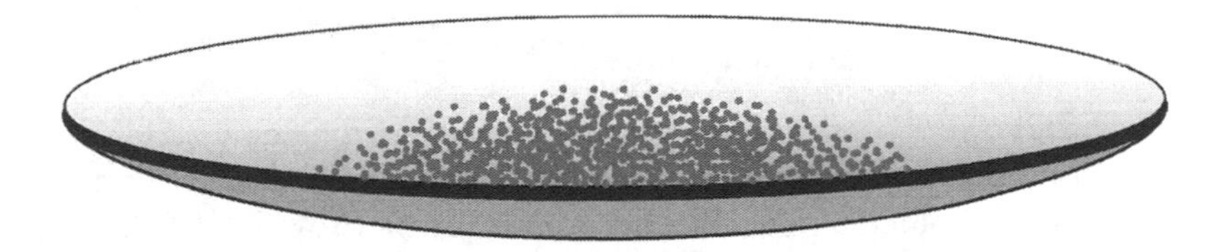

FIGURE 7.3
Watch glass. A watch glass is a shallow dish with large surface area. It is usually used for drying powder samples in ambient atmosphere or in a dry oven.

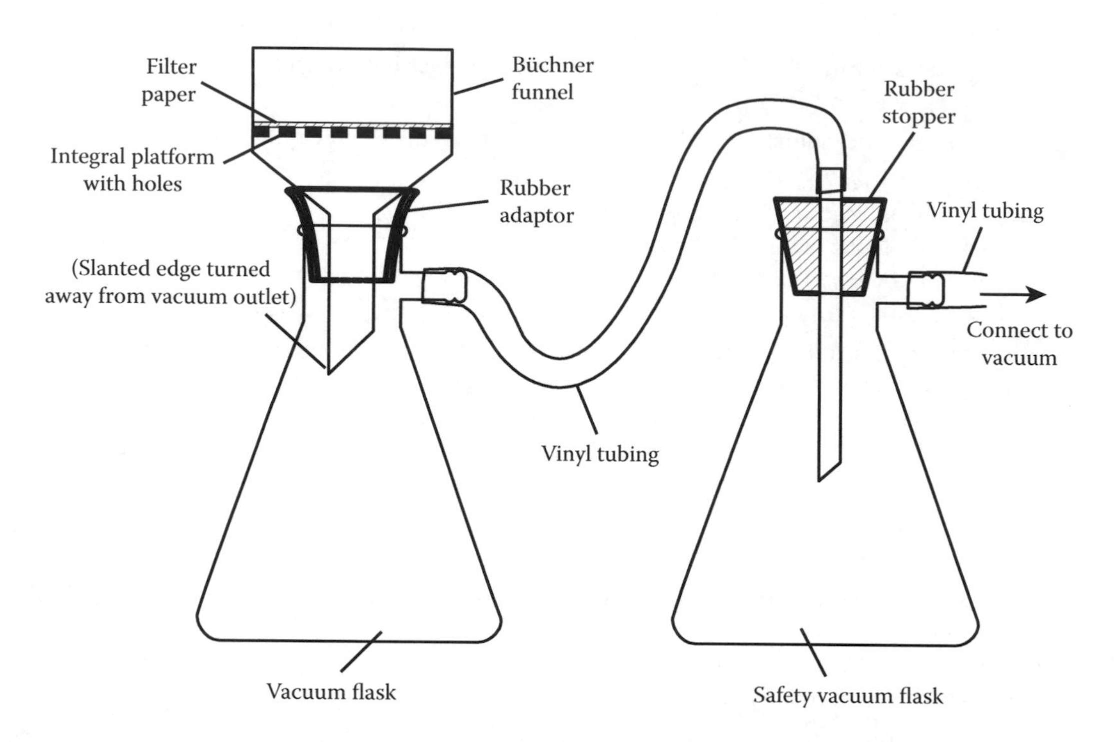

FIGURE 7.4
Setup for filtration using a Büchner funnel. A water aspirator pump can be used as a convenient source of vacuum. Press down the funnel when turning on the vacuum until it is held down by the vacuum. Handle ceramic Büchner funnels with care to avoid breakage.

2. (Optional) *Centrifugation*
   a. *Note:* The centrifugation step helps to reduce the amount of time needed for the filtration step by removing the small particles that can clog the filter paper.
   b. Stir your group's apatite reaction mixture at medium speed for 1 minute, and then pour it into the two 50-ml conical centrifuge tubes. Balance the tubes.
   c. Centrifuge the reaction mixture at 3000 rpm for 5 minutes. Discard the supernatant. (Note that the supernatant might appear milky since it contains very small HA/FA particles.)
   d. Pour de-ionized water into the two tubes to about half full. Break loose the pellet with a plastic stirring rod, and add water to near full. Balance the tubes, cap the tubes tightly, and then shake them vigorously (by hand or on a vortexer) to resuspend the pellets.
   e. Repeat the last two steps twice more.
3. *Filtration*
   a. Set up the Büchner funnel with filtration flask using the rubber adaptor (see Figure 7.4), and place a piece of filter paper in the funnel. Wet the paper with de-ionized water. Turn up the aspirator pump and press down the funnel until it is held in place through vacuum suctioning.
   b. At this point you should have the HA/FA suspension in two 50-ml tubes. Swirl the apatite suspension before pouring it onto the filter paper.

c. Wait until the liquid disappears from the surface of the solid, and then rinse the solid by adding just enough water to cover the surface. Repeat the rinsing twice more with water, once with 70% ethanol and then once with 100% ethanol.

d. After the last rinse, let aspiration continue for at least 30 minutes. When finished, *first* disconnect the tubing from the vacuum flask, and then shut off the water (or the tap water might rush into the flask because it still holds a vacuum).

4. *Drying:* Take down the Büchner funnel. Observe the collected HA (or FA). Dig the filter paper loose with a spatula and transfer the filtrate onto a watch glass; it is not necessary to remove the filter paper at this point. Dry the HA/FA in a dry oven with temperature set between 100–120°C overnight.

5. *Finishing up:* Empty the vacuum flask. Wash off any solid on the rubber adaptor. Wash and rinse the Büchner funnel and the spatulas, and air-dry on paper towels. (Do not hang up the Büchner funnel. It is heavy and will break if dropped.)

Fixed-Angle Rotor vs. Swinging Bucket Rotor for Centrifugation

A fixed-angle rotor has a solid body with slots oriented at a fixed angle to hold centrifuge tubes (Figure 7.5a). As the centrifuge tubes rotate, the centrifugal force, the "g-force," is perpendicular to the rotational axis, thus the precipitate accumulates at the outermost point near the bottom of the tubes. For a typical centrifuge, fixed-angle rotors usually can run at higher rotational speed than swinging bucket rotor. Fixed-angle rotors are appropriate for solids that require high g-force to precipitate. On the other hand, the precipitant must be able to sufficiently adhere to itself and form a pellet on the side near the bottom of the centrifuge tube.

For a swinging bucket rotor (Figure 7.5b), the centrifuge tubes and the tube holders ("buckets") are attached to the drive shaft with hinges. When still, the buckets are parallel to the rotational axis, but as the rotational speed of the rotor increases, the buckets are swung higher and higher until they are almost perpendicular to the rotational axis. The precipitant is then driven to the bottom of the centrifuge tubes. Although operating at lower rotational speed, the swinging bucket rotor is ideal for loose particulate materials. Notice that when we centrifuge cells (in Modules III and IV), we also use the swinging bucket mechanism.

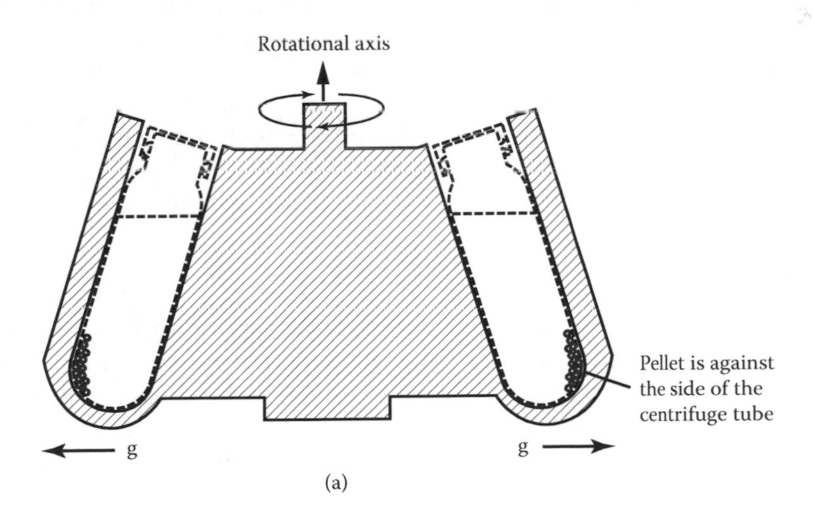

FIGURE 7.5
Fixed-angle rotor (a) versus swinging bucket rotor.

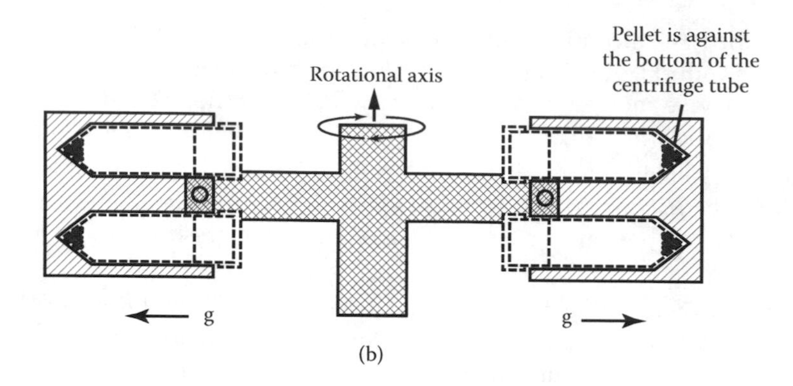

FIGURE 7.5
Fixed-angle rotor (b) for centrifugation.

Session 2. X-Ray Powder Diffraction of HA and FA

In synthesis of HA and FA using wet precipitation, the successful formation of HA/FA depends on the correct stoichiometric ratio of the reactants, the right pH, the reaction time, and the temperature. However, the starting material, calcium hydroxide, and the intermediates all appear as white powders, so how do we know for sure that we have actually synthesized HA/FA? The answer is x-ray powder diffraction.

X-ray diffraction is the basis of a suite of powerful techniques for structural analyses of materials. A full description is beyond the scope of this text and can be found in textbooks dedicated to fundaments of x-ray crystallography and diffraction (Hammond 2001). A "powder" sample of a given compound is defined as one that consists of crystallites of microscopic sizes. The average size of these crystallites is defined as the "grain size" (Figure 7.6A). When an x-ray beam hits a powder sample, it is diffracted by the crystal lattice of the crystallites (Figure 7.6B). Since x-rays are waves, the diffracted rays interact with each other in the form of wave interference; constructive interference takes place when two waves are in phase and results in peaks with higher intensities than the original waves, whereas destructive interference results in regions of no significant diffracted intensity. In other words, in order for the diffracted x-rays to be observed, constructive interference must take place. As shown in Figure 7.6B, when rays 1 and 2 in the incoming x-rays hit the lattice, the path for ray 2 is longer by $2d\cdot\sin\theta$. In order for the two diffracted rays to be in phase, which is required for constructive interference, this path length difference, $2d\cdot\sin\theta$, must be an integral number of the x-ray wavelength, or

$$n\lambda = 2d\sin\theta \qquad \text{(MV.1)}$$

where n is an integer, λ is the wavelength of the x-ray, d is the distance between the lattice planes, and θ is the incidence angle of the incoming x-rays with respect to the lattice planes. This relation is known as Bragg's law, first described by W. L. Bragg and W. H. Bragg.

In a typical powder diffraction experiment, you will do what is known as a $\theta - 2\theta$ scan. The sample is rotated through an angle θ, while the detector is rotated through an angle 2θ. Diffracted x-ray intensities are measured at the 2θ angle, known as the diffraction angle, and the result is usually plotted as a profile of intensity vs. 2θ. Therefore, for the example depicted in Figure 7.6B, a peak would appear in the diffraction profile because constructive

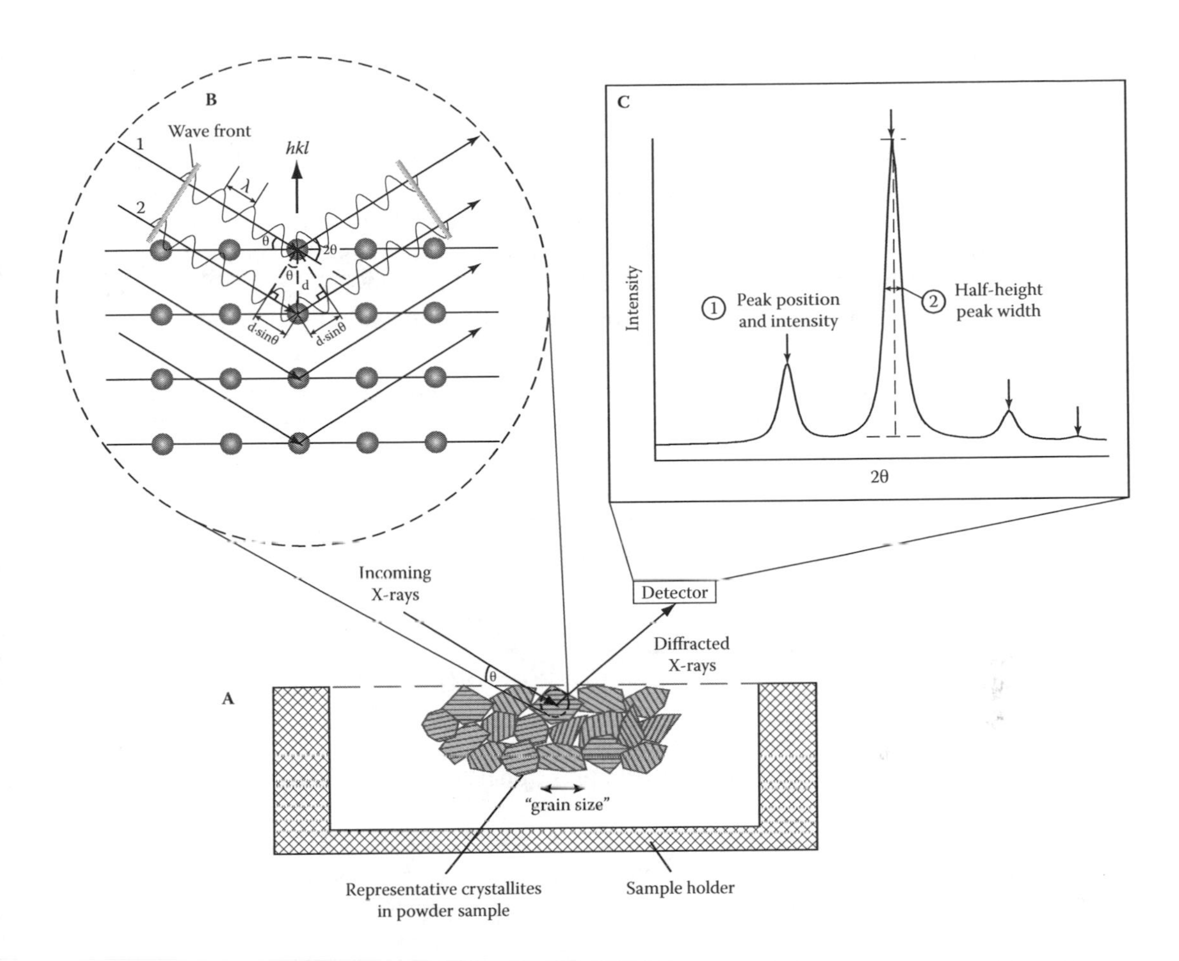

FIGURE 7.6
X-ray powder diffraction. (A) A powder sample consists of randomly oriented crystallites with a "grain size" or the average size of the crystallites expressed as a length. The powder sample is packed into a sample holder and made flush with the sample holder surface; this allows accurate measurement of the incident angle, θ, of the incoming x-rays. The diffracted x-ray intensity is measured by a detector set at an angle of 2θ with regard to the incident beam. (B) Rays 1 and 2 are diffracted off two lattice planes defined by a given set lattice indices (*hkl*). After diffraction, the wave front of ray 2 has traveled $2d{\cdot}\sin\theta$ longer than ray 1. When $2d{\cdot}\sin\theta$ is exactly two times of the wavelength λ, the diffracted rays are in phase. This condition satisfies Bragg's law, and constructive interference occurs as a result. (C) In an x-ray powder diffraction experiment, the diffracted x-ray is measured as a function of the increasing incident angle θ. The diffraction result is a profile of x-ray intensity vs. 2θ. Among others, two types of information can be obtained: (1) the peak positions and intensities, which form the "fingerprints" that can be used to identify the compound(s) in the sample, and (2) the linewidth, which correlates with the grain size of the crystallites, and serves as an indication for the crystallinity of the sample.

interference occurs at this particular angle; another peak would appear at the next θ angle that satisfies Bragg's law, and so on. Each diffraction plane is distinguished by a set of indices, (*hkl*), which defines its spatial orientation. Thus each Bragg peak is associated with a set of *hkl* values. An actual compound will have a more complicated crystal lattice than the simple one in Figure 7.6B, but the same diffraction principles apply. Measured diffraction profiles reflect the lattice spacing and the atoms that make up the lattice, from which a lot of structural information about the sample can be extracted. For our experiments we will focus on two types of information (Figure 7.6C):

1. The positions and intensities of the peaks: A crystal lattice can be divided into repeating units, called unit cells (and we have already seen the unit cell of HA in Figure 7.2). Different unit cells (composed of different spatial arrangement of atoms) will give different diffraction profiles, akin to fingerprint profiles. Diffraction profiles of tens of thousands of compounds have been measured and/or calculated and stored in databases; an unknown compound can be identified by searching for a match of its diffraction profile in such databases. This can be done even if the sample is a mixture of several compounds—the "fingerprints" of the different compounds will simply appear together in the diffraction profile and can be easily resolved.
2. The linewidth: The linewidth of a diffraction profile is characterized by the half-height peak width, and is related to the crystallinity of the sample. Crystallinity in turn depends on the "grain size" of the crystallites the sample consists of: a highly crystalline sample has crystallites with large "grain size," and a sample with smaller crystallites is said to be less crystalline. The size of the crystallites can be calculated using Scherrer's equation:

$$\text{Crystallite Size} = \frac{k \times \lambda}{\cos(\theta) \times (HHPW)} \quad \text{(MV.2)}$$

in which κ is the associated lattice index, λ is the wavelength of the x-ray, θ is the incident angle, and *HHPW* is the half-height peak width.

In this session we will obtain x-ray diffraction (XRD) profiles of HA and FA. The presence of these compounds can be confirmed by comparing the measured XRD profiles to the standard profiles. Also, we will study the linewidths of XRD peaks to obtain information on the samples' crystallinity.

Procedures

Part 1 (Day 1). Thermal Treatment of HA/FA

1. *Check-in*
   a. Samples and materials
      - Your group's dried HA (or FA) powders
   b. Special equipment and supplies
      - 100-ml porcelain crucible
      - Sintering oven, temperature range ≥600°C
2. *Thermal treatment*
   a. Weigh your group's HA (or FA) powder.
   b. Transfer about half of the HA (or FA) powder in the 100-ml porcelain crucible and place the crucible in the sintering oven. Once samples of all groups are in place, your instructor will program the temperature of the oven to increase at the rate of 2.5°C/minute and hold at 500°C for 5 hours.
   c. Store and label the rest of the HA (or FA) powder in a vial, a tube, or a bottle.

Part 2 (Day 2). X-Ray Powder Diffraction

1. *Check-in*
   a. Samples and materials
      - Thermally treated HA/FA
      - Untreated HA/FA
   b. Special equipment and supplies
      - Mortar and pestle
      - (Optional) Ball mill
      - X-Ray diffractometer and sample holders
2. *Milling the powder samples:* Grind the HA or FA samples with a ball mill or with the mortar and pestle until no chunks or clumps are visible. Clean the ball mill in between samples according to the manufacturer's instructions. For the mortar and pestle, simply wipe off the residual powder with Kimwipe tissue paper.
3. *Packing the samples into sample holders*
   a. *Note:* Tight and even compaction of the powder sample is very important for obtaining good quality x-ray diffraction data. Configuration of sample holder varies, but it usually includes a "window" to be filled with the powder sample. A key to good powder compaction is to totally fill up the sample window and not leave any voids.
   b. With a spatula, add and spread sample powder to the window until it is filled and brimmed with loose powder. With a glass slide (or an object that has a smooth, flat surface), press down the powder, and then use the edge of the slide to level the sample surface. Repeat this process until the surface of the sample is smooth and flush with the rest of the sample holder surface, and the sample is sufficiently compact that it stays in place when the sample holder is held up vertically. Wipe away any powder left outside of the sample window.
4. *Running the experiment:* Set up an x-ray powder diffraction measurement on the x-ray diffractometer according to the instrument's user manual and the following guidelines:
   a. Each group should run at least one sample of untreated HA or FA and its thermally treated counterpart.
   b. For CuKα x-rays, the minimal range for the diffraction angle 2θ should be 20–60°. This range contains most of the "fingerprint" peaks for apatite structures.
   c. It is recommended that the scanning rate be set to <2.0°/minute for typical diffractometers based on fixed-tube sources. The scanning rate will affect the quality of the data: a slower scanning rate will result in higher signal-to-noise ratios but will require longer data collection time. Since the HA/FA produced through wet precipitation is expected to have a low degree of crystallinity, it is important to scan at a sufficiently slow rate to obtain high-quality data.
   d. After obtaining the data, search and find peaks in the resulting diffraction profiles. (You can also manually measure the 2θ values of the peaks if you do not have access to software that can search for peaks automatically.)

e. Make sure that you have the following information in your lab notebook: the make/model of the x-ray diffractometer, a brief operation manual, the parameters for running each experiment, and the filenames and hard disk location of the results.

Data Processing

1. Use the calculated x-ray powder diffraction profile of HA (Figure 7.7 and Table 7.1) as a reference to identify and label the HA peaks in your own diffraction profiles. Use a research publication (Santos et al. 2004) as an example for proper labeling and presentation of your data. *Note:*
   a. Many factors can influence the actual crystal structure of the HA/FA that you have synthesized; thus, the peak positions and intensities might not match the values in Table 7.1 exactly, but the measured peak positions should be within ± 0.1° of the calculated values.
   b. The crystal structures of FA and HA are almost identical, with the difference of F^- substitution of the OH^- in the unit cells. The peak position in Table 7.1 is applicable to FA as well.
2. Use a software program (such as JADE from Materials Data) to measure the linewidths of your XRD data and calculate the grain sizes of your samples. If such software is not available, you can still quantitatively compare the widths of the peaks between untreated HA/FA and thermally treated HA/FA, and infer the relative degree of crystallinity of the samples. Simply fit the peaks with a Lorentzian peak shape and extract the full width at half maximum (FWHM).

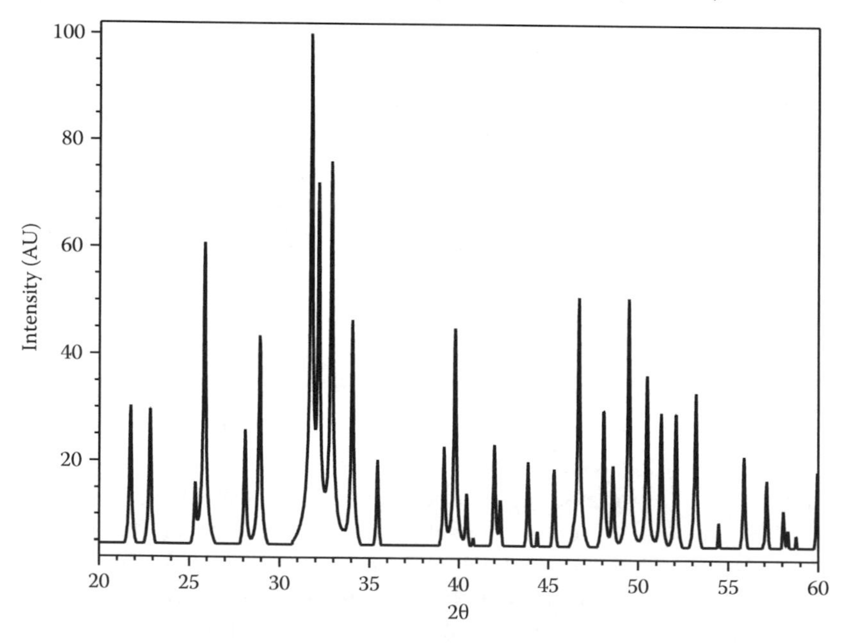

FIGURE 7.7
A calculated x-ray powder diffraction profile of HA. It is calculated based on a structure of HA (Leventouria et al. 2003) using the software Mercury (Macrae et al. 2008) from Cambridge Crystallographic Data Centre.

TABLE 7.1

Peak Positions and Relative Intensities of a Calculated HA XRD Profile (See Figure 7)

2θ	Relative Intensity	2θ	Relative Intensity
21.780	30.244	44.400	7.1818
22.880	29.638	45.340	18.899
25.380	15.993	46.720	50.979
25.880	60.736	48.100	29.934
28.140	25.937	48.620	19.640
28.940	43.439	49.500	50.853
31.780	**100.00**	50.520	36.537
32.200	72.085	51.300	29.533
32.920	76.009	52.120	29.404
34.080	46.454	53.220	33.282
35.480	20.396	54.480	9.0574
39.220	23.025	55.900	21.473
39.820	45.049	57.160	17.050
40.460	14.198	58.080	11.378
40.840	5.8475	58.340	7.6803
42.020	23.389	58.780	6.6567
42.340	12.986	59.980	18.671
43.880	20.237		

Session 3. Fabrication of Porous HA/FA-PLGA Composites Using a Salt-Leaching Method

Bone is a composite of HA and collagen. Structurally, the mineral phase provides mechanical stiffness and the collagen phase provides ductility and mechanical toughness. The combination of these two properties results in the unique qualities of bone: extremely hard and strong, yet tolerant to small degree of deformation, so that it is not easily shattered. (Many materials are extremely hard but also extremely brittle, which would not be a good property for bone.) In the search for bone-replacement materials, researchers want to mimic bone and take advantages of the properties offered by compositing minerals and organics. Various forms of HA composites have been studied. HA-collagen composite is a natural choice because it mimics the natural composition of bone, but collagen is an expensive material, and non-human collagen can be antigenic to some individuals. On the other hand, HA-polymer composites offer stronger mechanical properties, lower cost, and "tunable" biodegradation behavior. Commercial products based on HA-PLGA composite are already in clinical use.

Porosity is another important factor for bone-replacement materials besides strength. Interconnected pores enable bone cell migration, nutrient exchange, vascularization, and eventual new bone ingrowth. Porosity is usually characterized by two parameters: the pore size and the porosity percentage. To support bone ingrowth, the pore size should be >100 μm, but smaller pores are important for nutrient exchange and cell adhesion; thus, a polydisperse distribution for pore size is usually desirable. The porosity percentage is

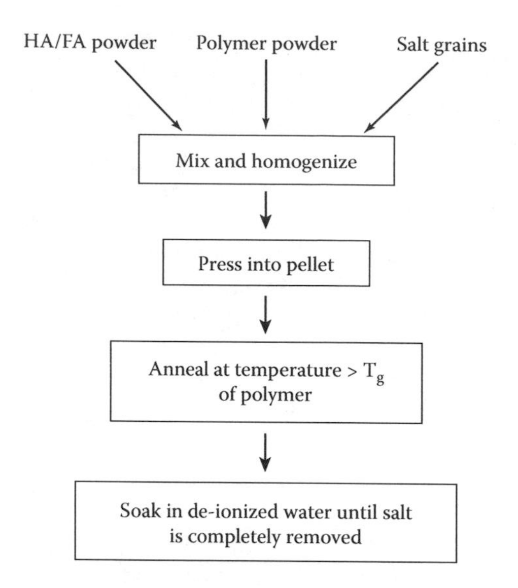

FIGURE 7.8
Procedures for fabricating porous of HA/FA polymer composites using a salt-leaching method.

expressed as the sum of the void volumes of the pores as a percentage of the total volume of the material. A higher degree of porosity promotes bone ingrowth but generally reduces the mechanical strength of the material. It is partly because of this that porous HA-PLGA composites are generally used for nonstructural bone-replacement applications such as bone void filling.

In this session we will fabricate porous HA-PLGA composites. We will use a salt-leaching method (Figure 7.8), in which salt grains are used as a space-filling material that can later be dissolved by water to create pores. The degree of porosity and the interconnectivity of the pores can be controlled by the weight percentage of the salt, and the pore size is related to the grain size of the salt.

Procedures

1. *Check-in*
   a. Samples and materials
      - HA and FA powder
      - PLGA powder
      - NaCl powder or grains (preferably milled or ground to grain size of 100–500 μm)
   b. Special equipment and supplies
      - Mortar and pestle
      - Aluminum foil
      - Lubricant
      - Hydraulic pellet press
      - Stainless steel pellet die, 0.5″ or 0.75″ in diameter
      - Dry oven set to 100°C

TABLE 7.2
An Example of Sample Fabrication Assignment

Group	Composites	Control
A[a]	HA-P[b] #1 and #2	P
B	HA-P #1 and #2	P
C	FA-P #1 and #2	P
D	FA-P #1 and #2	P

[a] This is an example for a class with four groups, A, B, C, and D.
[b] HA-P, HA-polymer composite; FA-P, FA-polymer composite; P, polymer only.

2. *Fabricating HA/FA-PLGA-NaCl pellets*
   a. *Note:* Control samples are fabricated using the same procedures as those for HA-PLGA composites, but without the HA. (Other polymers can be used in substitution for PLGA.)
   b. Weigh 1.30 g NaCl, 0.40 g PLGA powder, and 0.30 g HA or FA powder. For the control, weigh 1.30 g NaCl and 0.70 g PLGA. *Note:* Each group should prepare at least two HA/FA-PLGA-NaCl pellets and one PLGA-NaCl pellet. (See Table 7.2 for an example of fabrication assignment for a class with four student groups.)
   c. Add the above ingredients to the mortar and grind ~10 times using the pestle. The resulting powder should look homogenous, with an appearance of cake mix. Use a plastic spatula to collect the powder to a small beaker or weighing boat. Be sure to label the samples. Use Kimwipe tissues to wipe the mortar and pestle clean after use.
   d. Add the mixture to the pellet die and press it into a pellet using the hydraulic pellet press. The target force is 5000 lbs; hold the pressure for 2 minutes before releasing. Make a duplicate pellet using the same procedures.
3. *Heat incubation:* Rub a thin, almost invisible layer of lubricant or machine oil on a ~2″ × 2″ aluminum foil. (Excessive oil will soil the samples.) Place the pellets on the foil; make sure that the pellets are not in contact with each other (or they might fuse together). Incubate the pellet at ~100°C for 15 minutes. Remove the pellets from the oven and allow them to cool for 10 minutes. *Note:* If a polymer other than PLGA is used, be sure to adjust the oven temperature to greater than the T_g of the polymer.
4. *Salt leaching:* After the pellets are cooled, use a permanent marker to label each with "HA" or "FA" and your group's name. (You can write directly onto the pellets.) Place the pellets in a collective beaker, and add de-ionized water to dissolve (or "leach out") the NaCl. Your instructors will change the water daily until the next session. Alternatively, a drip-leaching device (Figure 7.9) that attaches to a water tank can be used to continuously remove the leached salt.

Session 4. Characterization of the Porous HA/FA-PLGA Composites—Acid Degradation and Compression Testing

In Session 3 we fabricated porous HA/FA-PLGA composites using a salt-leaching method. Are these composites suitable as bone-replacement materials? The answer is

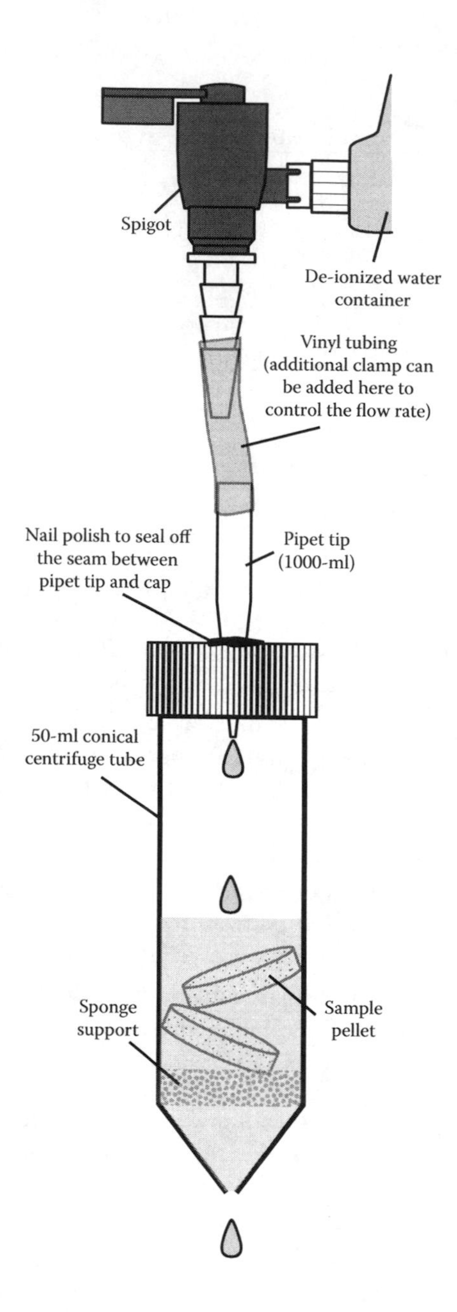

FIGURE 7.9
A drip-leaching device. This device works by constantly removing salt from the system while providing a steady replenishment of fresh water to the system. It can be made from simple supplies normally found in a lab: a 50-ml conical centrifuge tube, a pipette tip, a piece of vinyl tubing, and a de-ionized water container with a spigot. First drill a small hole in the cap of the 50-ml tube, then insert the pipette tip, and then seal the seam with nail polish (basically, epoxy glue) or other suitable glue. Also puncture a small hole at the bottom of the 50-ml tube. To put it in use, first temporarily seal off the bottom hole (with a Parafilm plug, for example), place a piece of sponge on the bottom of the tube as support, and then add de-ionized water into the tube to half full. Place the samples to be salt-leached inside. Screw on the cap tightly, and then attach the tube to the spigot of the water container via the pipette tip and the vinyl tubing. Open the spigot slightly, and remove the temporary plug at the bottom of the tube. Water should start to drip out. Adjust the spigot to control the flow rate. Roughly measure the flow rate with a timer and any graduated tube. It should be <1 ml/min. Allow the leaching to proceed for >48 hours.

not simple and will rely on characterizations of these composites in many different aspects, including porosity, biocompatibility, mechanical properties, and biodegradability. For example, to test the biocompatibility of the porous composites, we can use a cell proliferation test similar to the one we used in Module III (Chapter 5): we can seed osteoblasts into a scaffold made with a porous HA-PLGA composite and observe the proliferation of the cells. In this session we will characterize our porous HA/FA-PLGA composites using a combination of *in vitro* acid degradation and compression testing. For bone-replacement materials, *in vivo* degradation of the material is often concomitant with new bone growth, and it takes place in a physiological environment with near-neutral pH. However, the time scale for *in vivo* degradation could be weeks or months if not years. Although *in vivo* studies are ultimately necessary, we can begin by studying the degradation *in vitro* using acid degradation: The material is placed in an acidic solution, and then observed for the progression of degradation and its effect on the physical or chemical properties of the material. In a way, *in vitro* acid degradation can be viewed as an accelerated process that mimics the *in vivo* degradation. (This strategy is frequently used in materials testing.) Another study that we can conduct is a comparison of chemical stability between HA and FA. It is known that FA is more chemically stable than HA, but will it affect the acid degradation of the composites? Or how much effect will it have? These are issues that can be addressed by our experiments.

As acid degradation progresses, we expect the mechanical properties of the samples to be weakened, but we need to be able to measure such changes. The degradation process can be monitored in different ways. For example, the weight of a given sample can be measured to monitor the mass lost due to degradation, or the microscopic morphology of the sample can be observed to study material erosion due to degradation. In this session we use compression testing to measure the mechanical properties of the samples. Naturally, compressive properties are critical features for any bone-replacement material since a major function of bone is load-bearing. One of the most important mechanical properties for materials is the Young's modulus, a parameter that describes the elastic, or linear, response of the material to external force according to Hook's law; it quantifies the *stiffness* of a given material. Compressive Young's modulus E is obtained by measuring compressive stress σ vs. strain ε:

$$E = \frac{\sigma}{\varepsilon} \tag{MV.3}$$

Stress measures force per cross-section area (Figure 7.10),

$$\sigma = \frac{F}{A} \tag{MV.4}$$

and strain measures the deformation of a material under stress,

$$\varepsilon = \frac{\Delta l}{l_0} \tag{MV.5}$$

In Figure 7.10,

$$\Delta l = \Delta l_1 + \Delta l_2$$

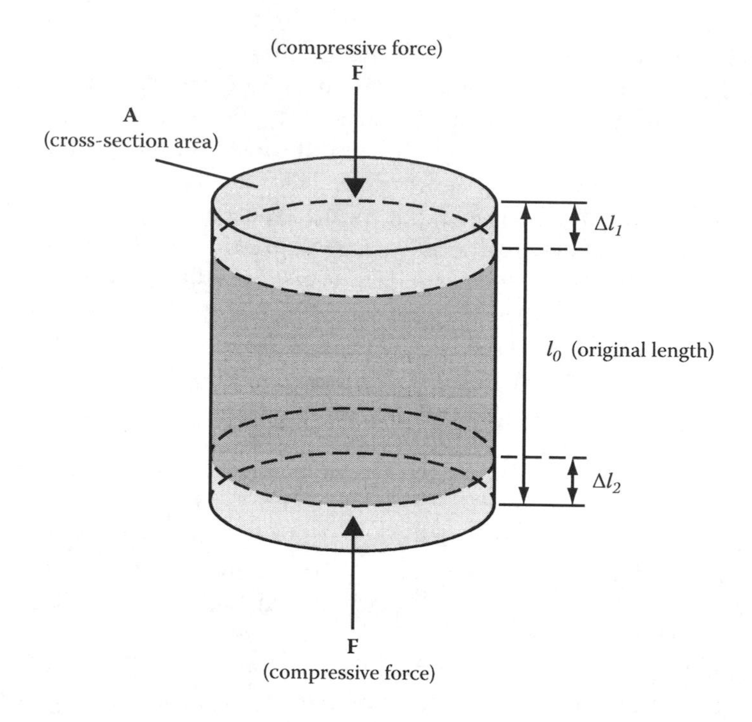

FIGURE 7.10
Deformation of materials under compression.

Another important mechanical property is the ultimate strength. In a compression testing, the ultimate strength is the compressive stress that a material can endure before structural failure. No material can bear an unlimited amount of load without failure, and the ultimate strength sets the upper limit for a mechanical application of the material.

Procedures

Part 1 (Day 1). In Vitro *Acid Degradation of the Composites*

1. *Check-in*
   a. Samples and materials
      - Porous HA/FA-PLGA composite pellets (from Session 3)
   b. Reagents
      - pH 4.01 buffer (calibration solution)
      - PBS
   c. Special equipment and supplies
      - Beakers, 50 ml or 100 ml
2. *Sample treatment*
   a. *Note:* In order to make a systematic study of the degradation behaviors of HA/FA-PLGA composites, we need to compare the HA-PLGA composite and the FA-PLGA composite before and after acid degradation, and use the PLGA-only pellets as controls. See Table 7.3 for an example of an acid

TABLE 7.3

An Example of Sample Degradation Treatment Scheme and Group Assignment

HA-Polymer Composites		FA-Polymer Composites		Control (Polymer Only)	
Acid	PBS	Acid	PBS	Acid	PBS
A[a]: HA-P[b] #1	A: HA-P #2	B: FA-P #1	B: FA-P #2	A: P	B: P
C: HA-P #1	C: HA-P #2	D: FA-P #1	D: FA-P #2	C: P	D: P

[a] The group names correspond to the example in Table 7.2.
[b] HA-P, HA-polymer composite; FA-P, FA-polymer composite; P, polymer only.

degradation scheme and assignment. PBS is used as a control solution for the acidic solution.

b. Label one beaker with "acid" and another with "PBS" along with your group's name and today's date. Add the pH 4.01 buffer solution and PBS in the respective beaker according to a ratio of 1 ml solution for every 10 mg of HA (or FA) in the sample. (Here we use the pH 4.01 calibration solution as a weak acid solution.) Place your group's porous HA/FA-PLGA composite pellets in the corresponding solution, and the control sample in one of the solutions according to your group's assignment. Seal the beakers with Parafilm and place them in a 37°C incubator for at least 48 hours.

3. *Postdegradation treatment [by your instructor]:* After the designated period of time for the *in vitro* acid degradation, your instructor will remove all the samples from the solutions, rinse them with de-ionized water thoroughly, and let them dry overnight in a dry vacuum or in a ventilated hood.

Part 2 (Day 2). Compression Testing

1. *Check-in*
   a. Samples and materials
      - Acid-degraded and nondegraded HA/FA-PLGA composites, rinsed and dried
      - Acid-treated and PBS-treated porous polymer samples, rinsed and dried
   b. Special equipment and supplies
      - Micrometer
      - Compression testing instrument
2. *Sample measurement:* Use the micrometer to measure the diameter and thickness of the sample pellets. Make at least four repeated measurements for each parameter and calculate the average. The precision of the measurements should be 0.1 mm.
3. *Compression testing:* Set up compression testing according to the instrument's user manual and the following guidelines:
   a. ***Safety note:*** Under normal conditions, it is unlikely that the compression will produce projectile fragments. However, a protection shield should be in place after loading the sample and should not be removed until the compression testing has been terminated.
   b. Make sure that the instrument has been calibrated.

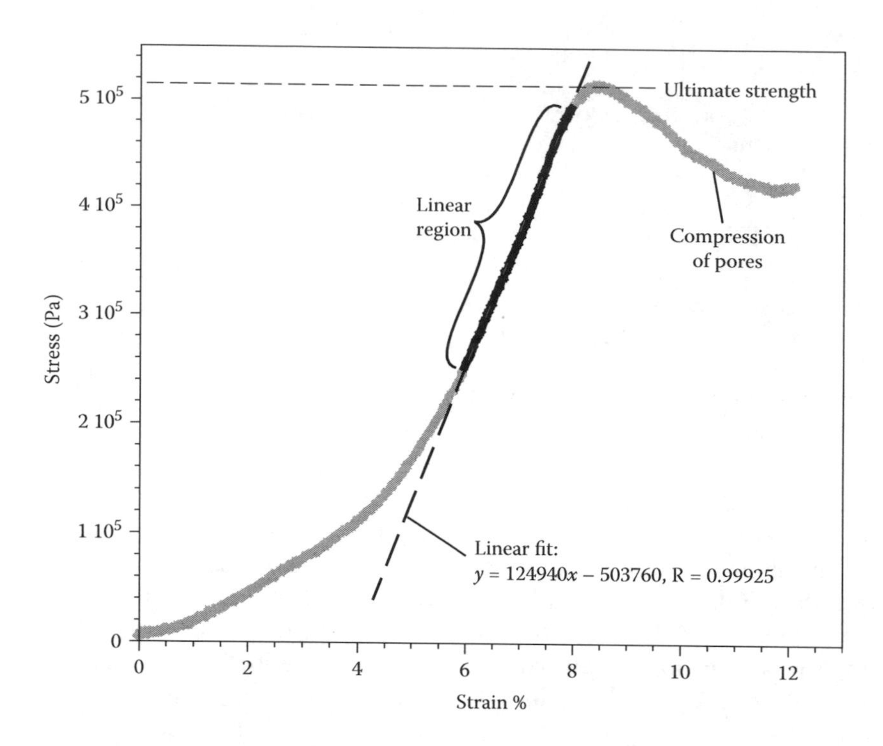

FIGURE 7.11
Example of a stress–strain curve. Data were obtained from a compression test of a porous HA-PLGA composite sample. Notice that the strain data were converted to % for the x axis. To obtain the compressive modulus, a linear zone has been selected *before the stress reaches the ultimate point,* and fitted with a linear function. The slope of the linear function, 124940, is then used to calculate the compressive modulus, which is 12.5 MPa. The ultimate strength corresponds to the highest point of stress before it starts to decrease, signaling a breakdown of the material structure. The material can be further compressed after this point, likely through the collapse of the pores.

c. Many compression-testing instruments also measure extension; make sure that the instrument (and the computer) is set for compression.
d. The compression speed (or displacement rate) should be 0.0010 mm/sec or similar.
e. The geometry of the pellet samples should be set to "circular" or similar to reflect to cylindrical shape of the pellets; the measured diameter and length should be entered in the computer individually for each pellet before the test.
f. If possible, set up the software that controls the instrument so that it automatically calculates the stress and strain.
g. The testing can be terminated after the force reaches the ultimate strength of the material. See the Data Processing section and Figure 7.11 for reference.
h. Make sure that you have the following information in your lab notebook: the make/model of the compression-testing instrument, a brief operation manual, parameters for running each experiment, a note of the temperature and the humidity, and the filenames and the hard disk location of the results.

Data Processing

1. Plot stress vs. strain from your experimental data. An example of a compression-testing profile of a similar porous HA-PLGA material is shown in Figure 7.11. *Note:*

a. The unit for stress should be Pascal.

b. Strain is dimensionless (or, it does not have a unit), but can be converted to % for the stress-strain plot. However, do not use the % values when calculating the modulus.

c. Because it is a compression testing, the displacement may be recorded as a negative number by some instrument. If that is the case, make sure to convert the data in order to obtain a normal-looking stress-strain plot.

2. Obtain the following parameters from the compression-testing curves:

   a. Compressive moduli: select a linear region just before the breaking point to calculate the compressive modulus for each testing curve (see Figure 7.11). *Note:* Do not use percentage values for the strain when calculating the modulus. For example, in Figure 7.10, the slope for the linear fit is 124940, and the modulus should be 12494000 Pa, i.e., 1.25×10^7 Pa, or 12.5 MPa.

   b. Ultimate strength: estimate the stress at the breaking point (see Figure 7.11).

3. Combine your data with all other groups in your class. For example, using the example given in Table 7.3, you can calculate the average and standard error of each sample treatment condition using duplicate data. For each sample, calculate the average and the standard error for the compressive modulus and the ultimate strength. Summarize the results in a table and illustrate them with bar graphs.

Questions

1. Which one(s) of the following statements regarding the synthesis and treatment of hydroxyapatite (HA) is(are) incorrect?

   a. "Amorphous" HA is more bioactive than crystalline HA.

   b. Sintering increases the crystallinity of HA.

   c. During wet precipitation, HA formed at a higher temperature has a higher crystallinity.

   d. At pH >10, when $Ca(OH)_2$ and H_3PO_4 are mixed at the right stoichiometric ratio, HA formation is instantaneously complete.

   e. When filtering HA in a Buchner funnel, ethanol is added to help to complete the chemical reaction.

2. In synthesis of HA using $Ca(OH)_2$ and H_3PO_4 as precursors, as $Ca(OH)_2$ is added to the H_3PO_4 solution, the pH increases. After the pH increases to a certain point, $Ca(OH)_2$ is no longer soluble, so how is it that the insoluble $Ca(OH)_2$ can react with phosphate to form HA?

3. A calcium phosphate-based bioceramics sample has two components: tricalcium phosphate (TCP), $Ca_3(PO_4)_2$, and HA, $Ca_{10}(PO_4)_6(OH)_2$. It is not clear how much of each component is present, but through elemental analysis, the Ca:P ratio of this sample is determined to be 1.60. What is the weight percentage of HA in this sample? The molecular weight for TCP is 310.18, and that for HA is 1004.62.

4. You are setting up an x-ray powder diffraction experiment. The 2θ range you want to scan is 10–70°, but you have only 30 minutes to complete the experiment. What scan rate should you use for the run?

5. In a given 2θ range, compound A has six diffraction peaks, and compound B has four diffraction peaks. (The peaks for compound A and compound B do not overlap.) You have a sample that is a 2:1 mixture of compound A and compound B. How many diffraction peaks will appear in this 2θ range if you run an XRD experiment on this sample?
6. When fabricating porous materials using the salt-leaching method, how do we control the porosity, and how do we control the pore size?
7. Polymer membrane can be prepared using a casting process: the polymer is first dissolved in a solvent, and the solution is cast onto an inert surface such as that of a glass plate; the solvent is then allowed to evaporate, after which a film is formed. Describe how you would prepare a porous membrane using a combination of casting and salt-leaching.
8. To estimate the compressive modulus of chicken tibia bone, a ~1 cm segment is isolated from the middle of a chicken drumstick bone, and the bone marrow is removed. The two ends of this bone segment are sanded with sandpaper until the surfaces are level and smooth (Figure 7.12A). To measure the cross-section area of this test specimen, several measurements are made for the inner diameters (i.d.) and the outer diameters (o.d.) due to the irregularity of the cross-section (Figure 7.12B). The average of the four o.d. measurements is 10.6 mm, and that for the four i.d. measurements is 6.9 mm. During the test, when the strain reaches 1.00%, the force is 552.63 Newton. What is the instantaneous stress and Young's modulus for this time point?
9. Table 7.4 summarizes the properties of some of the ceramics-polymer composite materials that have been developed by various research laboratories.
   a. Why is cortical bone included in the "dense composites" section, and cancellous bone included in the "porous composites" section for comparison?
   b. Study the properties of HA-PLGA composites (shaded) in Table 7.4. How does the porosity affect the stiffness of the material in general?
   c. Of the different types of composites in Question 9b, which one most closely resembles the composites that we have fabricated in our experiments?

Appendix. Recipes and Sources for Equipment, Reagents, and Supplies*

Session 1. Synthesis of HA and FA

- **Centrifuge with swinging bucket rotor.** See Module I appendix (Session 1). Swinging bucket rotor: Eppendorf cat. no. A-4-81.
- **Büchner funnel.** Coors porcelain Büchner funnel with fixed perforated plate, Fisher cat. no. 10-356D. *Sel. crit.:* Either porcelain or plastic funnels are OK.

* *Disclaimer*: Commercial sources for reagents listed are used as examples only. The listing does not represent endorsement by the author. Similar or comparable reagents can be purchased from other commercial sources. See selection criteria (*Sel. crit.*).

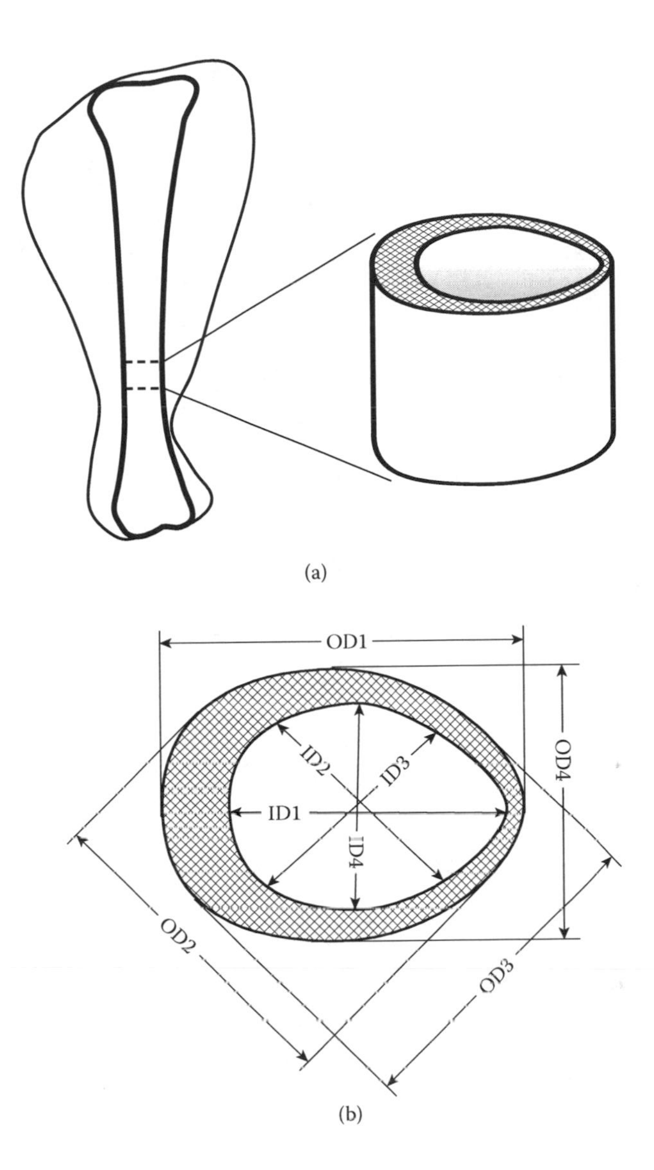

FIGURE 7.12
Estimating the modulus of chicken tibia bone. (a) A small segment of bone is removed from a chicken drumstick to be used as specimen for compression testing. (b) Measuring the cross-section area of the bone specimen. (See Question 8.)

- **Filter papers (diameter 90 mm).** Whatman qualitative grade plain circles, mfr. no. 1001-090, Fisher cat. no. 09-805D. *Sel. crit.:* Must fit the Büchner funnel.
- **Filtration flask.** Pyrex Vista heavy-wall filter flask Fisher cat. no. S76108B or S76108C. *Sel. crit.:* Similar volume.
- **Rubber funnel adaptor (for 1000-ml filtration flask).** Fisherbrand adapters for flask size 6, Fisher cat. no. 03-448-41. *Sel. crit.:* Must fit the filtration flask and the Büchner funnel.
- **Large watch glass.** Pyrex watch glass, mfr. no. C9985-100, Fisher cat. no. S34821. *Sel. crit.:* Diameter ~15 cm.

TABLE 7.4

Selection of Scaffold Composites Designed for Bone Tissue Engineering and Their Properties[a]

Scaffold Composite: Ceramic	Scaffold Composite: Polymer	Percentage of Ceramic (%)	Porosity (%)	Pore Size (μm)	Compressive (C), Tensile (t), Flexural (f) Strength (MPa)	Modulus (MPa)	Reference[b]
1. Dense Composites							
HA fiber[c]	PDLLA	2–10.5 (vol.)	—	—	45 (F)	$1.75–2.47 \times 10^3$	[149]
	PLLA	10–70 (wt.)	—	—	50–60 (F)	$6.4–12.8 \times 10^3$	[150]
HA	PLGA	40–85 (vol.)	—	—	22 (F)	1.1×10^3	[29,151]
β-TCP	PLLA-co-PEH	75 (wt.)	—	—	51 (F)	5.18×10^3	[152, 153]
	PPF	25 (wt.)	—	—	7.5–7.7 (C)	191–134	
A/W	PE	10–50 (vol.)	—	—	18–28 (B)	$0.9–5.7 \times 10^3$	[154]
Cortical bone					50–150 (T) 130–190 (C)	$12–18 \times 10^3$	[31, 74, 75]
2. Porous Composites							
Amorphous CaP	PLGA	28–75 (wt.)	75	>100		65	[28,155]
HA	PLLA	50 (wt.)	85–96	100–300	0.39 (C)	10–14	[127]
	PLGA	60–75 (wt.)	81–91	800–1800	0.07–0.22 (C)	2–7.5	[156]
	PLGA		30–40	110–150		337–1459	[157]
Bioglass®	PLGA	75 (wt.)	43	89	0.42 (C)	51	[93,148,158]
	PLLA	20–50 (wt.)	77–80	~100 (macro) ~10 (micro)	1.5–3.9 (T)	137–260	[17]
	PLGA	0.1–1 (wt.)		50–300			[15,148]
	PDLLA	5–29 (wt.)	94	~100 (macro) 10–50 (micro)	0.07–0.08	0.65–1.2	[132,134,136]
Phosphate glass	PLA-PDLLA	40 (wt.)	93–97				
A/W	PDLLA	20–40 (wt.)	85.5–95.2	98–154	0.017–0.020 (C)	0.075–0.12	[159,25]
Cancellous Bone					4–12 (C)	100–500	[107,108]

[a] Reprinted with permission from Elsevier, by Q. Rezwan, Z. Chen, J.J. Blaker, and A.R. Boccaccini "Biodegradable and bioactive porous polymer/inorganic composite scaffolds for bone tissue engineering," *Biomaterials*, 27, 3413–3431, 2006.

[b] References are cited in Rezwan et al. 2006.

[c] Abbreviations: HA, hydroxyapatite; TCP, tricalcium phosphate; A/W, apatite-wollastonite; CaP, calcium phosphate; PDLLA, poly-D,L-poly(lactic acid); PLLA, poly-L-poly(lactic acid); PLGA, poly(lactic-*co*-glycolic acid); PEH, poly(ethylenesebacate-*co*-hexamethylenesebacate); PPF, poly(propylene fumarate); PE, polyethylene; PLA, poly(lactic acid).

Session 2. X-Ray Powder Diffraction of HA and FA

- **100-ml porcelain crucible.** Coors crucible, high-form, Fisher cat. no. S32609-1CR; cover, cat. no. S32006CR. *Sel. crit.:* Withstand temperature >1000°C.

Session 3. Fabrication of Porous HA/FA-PLGA Composites Using a Salt-Leaching Method

- **PLGA powder.** Broken down from PLGA pellets with a ball mill under cryogenic conditions. Or alternatively, dissolve 5 g of PLGA pellets in 20 ml of dichloromethane, and then add the PLGA solution drop-wise to 150 ml of 1% PVA solution stirred at high speed. Allow the solvent to evaporate overnight. Afterward filter the suspension with a Büchner funnel, wash the PLGA powder with de-ionized water, and dry the PLGA powder at room temperature in air or with a vacuum.
- **Stainless steel pellet die (diameter 0.5″ or 0.75″).** MTI Corporation, cat. no. EQ-Die-18D or EQ-Die-12D. *Sel. crit.:* Best if the diameter is between 0.5″ and 0.75″.

References

de Leeuw, N., Local ordering of hydroxy groups in hydroxyapatite, *Chem. Commun.*, 1646–1647, 2001.

Farley, J., J. Wergedal, and D. Baylink, Fluoride directly stimulates proliferation and alkaline phosphatase activity of bone-forming cells, *Science*, 222, 330–332, 1983.

Hammond, C., *The basics of crystallography and diffraction*, 2nd ed., Oxford University Press, 2001.

Liu, C., Y. Huang, W. Shen, and J. Cui, Kinetics of hydroxyapatite precipitation at pH 10 to 11, *Biomaterials* 22, 301–306, 2001.

Santos, M.H., M. de Oliveira, L.P. de Freitas Souza, H.S. Mansur, and W.L. Vasconcelos, Synthesis control and characterization of hydroxyapatite prepared by wet precipitation process, *Mat. Res.*, 7, 625–630, 2004.

Vallet-Regí, M., Ceramics for medical applications, *J. Chem. Soc., Dalton Trans.*, 97–108, 2001.

Appendix: Answers

Module I

1. During sonication, the protein solution is broken up into microscopic droplets by the sonic pulses. These droplets scatter light, making the sonicated mixture opaque and giving it a milky appearance.
2. PVA is an amphiphilic surfactant. It can interact with both the hydrophilic aqueous phase and the hydrophobic organic phase, thus it serves to stabilize the interface between the dichloromethane microspheres and the water in the second-degree emulsion. Therefore, the presence of the 9% PVA solution is crucial to forming the second-degree emulsion, and de-ionized water cannot be used to substitute for the 9% PVA solution.
3. The success of a centrifugation run depends on the sedimentation speed of the particles, or the PLGA microspheres in this case, and the sedimentation speed in part depends on the density difference between the particles and the fluid, as well as the viscosity of the fluid. The 9% PVA solution that we used for the second-degree emulsion has relatively high density and viscosity compared to water, thus impeding sedimentation of the PLGA microspheres. By adding water, we can decrease both the density and viscosity of the PVA solution so that at a given centrifugation speed, sedimentation can complete within shorter centrifugation time.
4. Choice "c" is incorrect. PVA stabilizes the physical integrity of the droplets, but it is chemically inert. The time scale for PLGA degradation is months, so during the microsphere fabrication process, which lasts only several days, PLGA degradation is pretty much negligible.
5. Notice that the drug is dissolved in PBS, probably for better stability of the drug. The PBS is encapsulated along with the drug during the encapsulation process. The PBS that we use contains 155 *mM* NaCl, 1 *mM* KH_2PO_4, and 3 *mM* Na_2HPO_4, which means that the solid content for each liter of PBS is 14.68 g, or 14.68 mg/ml, and part of it is encapsulated along with the drug, giving an apparent yield of >100%.
6. In Session 1, we prepared the 50 mg/ml BSA solution by weighing ~50 mg of BSA. There are a number of problems with weighing: 1) the BSA could contain moisture; 2) the BSA might not be pure; and 3) when the BSA solid is dissolved, it increases the volume of the buffer. Therefore, the weight of the BSA solid is not suitable for determining the BSA concentration. Although we can use weighing to reach the target BSA concentration, we need a more accurate method to measure the actual concentration.

7. The drug load is

$$\text{D.L.} = \frac{\text{weight of encapsulated drug}}{\text{weight of microspheres}}$$

$$= \frac{4.50}{10.00} = 45.0\%$$

The encapsulation efficiency is

$$\text{E.E.} = \frac{\text{weight of encapsulated drug}}{\text{total weight of drug}}$$

$$= \frac{4.50 \times \frac{160}{10}}{100} = 72\%$$

8. Mistakes happen in the lab. Some mistakes cannot be fixed so that you have to start the experiment over, some mistakes do not affect the outcome of the experiment much, and some other mistakes can be fixed. For the four scenarios:
   a. Scenario 1: Depends on how much more is added. The polymer concentration can affect the encapsulation efficiency and drug load, but a small change will not have a significant effect. So if 5.5 ml dichloromethane is added instead of 5 ml, for example, it is probably OK, but if 10 ml is added, it'd be better to make another PLGA solution.
   b. Scenario 2: The volume of PBS used to extract BSA must be exact in order to determine the weight of encapsulated BSA. In this case, since the volume cannot be determined, this experiment will have to be started over again from the beginning.
   c. Scenario 3: For the Bradford assay, the mixing ratio between the Bradford reagent and the protein standard solution needs to be optimized to obtain the most accurate measurement. When the protocol is optimized for 50 µl BSA standard solution, 5 µl will most likely be insufficient to generate enough color, and the result will be significantly affected. You need to start this experiment over.
   d. Scenario 4: The difference between 55 µl and 50 µl is tolerable. But you need to make sure that the volume 55 µl is used for all of the standard samples as well as the unknown samples, since in a Bradford assay (and in other similar assays), it is critical that the standard samples and the unknown samples be prepared and measured under identical conditions.
9. One of the advantages for the test tube protocol is its higher sensitivity and accuracy. In this protocol, the color reaction is stronger since more protein is used, thus the measurements have higher signal-to-noise ratio. However, this protocol is not very efficient since samples are handled one by one. Also, it requires a lot of protein, so it is not suitable if you only have a small amount of the protein. The most important advantage of the microplate protocol is its efficiency, since up to 96 samples can be measured in one run using 96-well microplate. Therefore, the microplate protocol is especially suitable for handling large numbers of samples.

On the other hand, this protocol requires microplate reader, and is less sensitive and less accurate than the test tube protocol, although the latter can be partly compensated by measuring samples in duplicates or triplicates. For Session 1, we used the test tube protocol mainly because the concentrations of our samples were within the optimal range for this protocol, and also because we did not have a lot of samples to measure. For Session 4, we had a large number of samples to measure, and the protein concentrations of our samples were low, so the microplate protocol was more appropriate.

10. The fabrication process can dramatically affect the protein release profile.
    a. See Figure 2.14.
    b. There could be a number of reasons. The most obvious one is the size difference between the *w/o/w* microspheres and the *s/o/w* microspheres. The *s/o/w* microspheres have a larger surface-area-to-weight ratio due to their smaller sizes, which facilitates diffusion of water into and insulin out of the microspheres. Another factor could be the distribution of insulin inside the microspheres. The *w/o/w* method allows insulin to be distributed in the PLGA matrix evenly, thus minimizing the initial burst. The *s/o/w* method, on the other hand, starts with insulin solid, which is distributed less evenly and tends toward the surface of the microspheres, resulting in bigger burst release of the insulin.
    c. The large initial burst from *s/o/w* microspheres could increase the insulin concentration to toxic level, whereas the concentration of released insulin could be too low to be effective for the *w/o/w* microspheres due to the slow release.
    d. The *s/o/w* and *w/o/w* microspheres can be mixed at various ratios to adjust the insulin release profile, so that the insulin concentration is at the right therapeutic level.

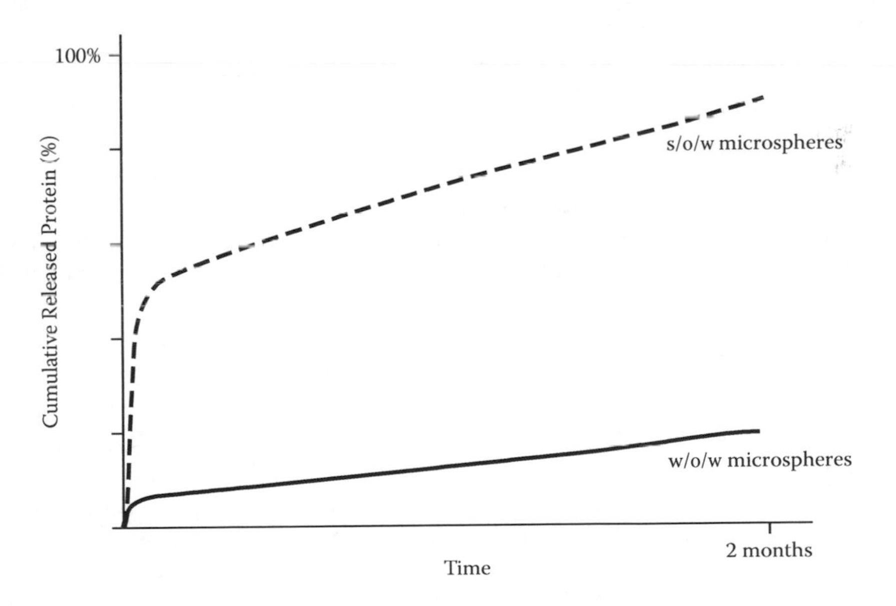

FIGURE 2.14
A sketch of two controlled release profiles. For the microspheres fabricated using w/o/w double emulsions, the initial burst is 5% and reaches 20% cumulative release in 2 months (solid line). For the microspheres fabricated using s/o/w emulsion, the initial burst is 50% and it reaches 90% cumulative release in 2 months (dotted line).

Module II

1. Choice "b" is correct. The aspect ratio of tropocollagen is ~3000/15, or ~200. Of the three choices, the aspect ratio for the pencil is ~19.3, for the hair is ~187.5, and for the flagpole is 120. Therefore, the one that is closest to collagen in shape (but not dimensions) is the hair. On a different but related note, the persistent length of type I tropocollagen is measured to be around 145–160 Å; thus, the molecule is quite flexible. However, this flexibility is significantly reduced when tropocollagen molecules assemble to form fibrils, and it is reduced further yet with cross-linking among the tropocollagen molecules.
2. The calf skin is freeze-fractured as a result of pulverization under cryogenic condition, increasing the surface areas for efficient extraction.
3. In fetal calf skin, a high percentage of the type I collagen is not yet cross-linked; therefore, we can get a higher yield of acid-soluble collagen. We can technically use mature bovine skin for the extraction. However, with a much lower yield, and as a material that is much tougher to handle, mature skin is not quite worth the effort (at least not for our experiments).
4. Choice "b" is correct: Tropocollagen is extracted. On the other hand, a small amount of tropocollagen dimers or trimers—i.e., groups of two or three tropocollagen molecules still linked together—might be present, too.
5. The extraction efficiency depends on the amount of solution available: More solution usually allows better extraction, although the concentration of the extract will likely decrease with increasing volume of the extraction solution, which can potentially affect the subsequent purification. However, a 10-ml difference out of a volume of 160 ml is not likely to make a significant difference. You should go on with the experiment.
6. When isolating a particular protein with salt precipitation, a precise salt concentration is critical, since solubility is an intrinsic property for a given protein. In this case, when the salt concentration is much higher than intended, it will likely reduce the solubility of proteins other than type I collagen, and these "contaminating" proteins will likely coprecipitate with type I collagen, thus defeating the purpose of the salt precipitation. If you only realize the mistake after you have obtained the precipitant, it is still not too late: you can just redissolve the precipitant to approximately the same volume as before, and repeat the salt-precipitation process with the *correct* amount of NaCl. You should check the purity of the target protein by running a SDS-PAGE of the checkpoint samples, of course.
7. Choice "c" is the best option. Blood plasma contains hundreds of proteins, but fibrinogen is present with quite high a concentration while most other proteins are present in low or even trace amounts. Therefore, salt precipitation is the most appropriate first step to 1) remove most of the other proteins, and 2) adjust the concentration of fibrinogen to a desired level for further purification. Dialysis in principle can remove molecules whose molecular weights are below the molecular weight cutoff (MWCO), but in practice this process is too inefficient. Gel electrophoresis is an analytical method and thus not suitable for large-scale preparation. Most proteins including fibrinogen cannot be extracted with acid; in

fact, most proteins will likely be denatured or deactivated in an acidic environment; collagen is a noted exception.

8. An easy way to change the salt composition of a protein solution is dialysis. In this case, you should dialyze your sample in a solution of 100 *mM* KCl and 50 *mM* phosphate.
9. Choice "c" is correct. The number of amino acids in a protein corresponds to its molecular weight, and SDS-PAGE resolves proteins by their molecular weights.
10. Rapid temperature decrease leads to rapid formation of ice crystals, but the ice crystals will be small because many nuclei are formed at the same time. Therefore, the sponge will have small pore size. On the other hand, slow freezing will result in large ice crystals and subsequently large pore size for the sponge.
11. An example for the similarity between the two methods: acetic acid is used to disrupt the collagen self-assembly, thus facilitating extraction or homogenization. An example for the difference between the two methods: When fabricating collagen sponges, the epidermis of the calf skin needs to be removed before further processing, whereas the whole calf skin is used for extraction of acid-soluble collagen.

Module III

1. Let's use the worksheet in Table 5.3 to work out a way to seed the T-25 flasks:

   First, let's determine the number of cells to seed for the three T-25 flasks:

   Total cell count at confluency for the three T-25 flasks: $N_{cnfl} = 3 \times 1.0 \times 10^6$

   We are seeding the cultures ahead for: $T = 3$ days = 72 hours

   The doubling time for NIH 3T3 is: $T_d = 24$ hours

   Number of cells to seed:

$$N_{seed} = N_{cnfl} \times 2^{-\frac{T}{T_d}} = 3.75 \times 10^5$$

   Next, let's determine the volume of the cells suspension that is needed for the seeding:

   Cell count of the cell suspension: C.C. = 5.0×10^5 ml^{-1}

   Volume of cell suspension needed: $V_{cs} = 3.75 \times 10^5 / 5.0 \times 10^5$ ml^{-1} = 0.75 ml

   Since V_{cs} <1.0 ml, we will make diluted cell suspension (DCS) by mixing 1.0 ml of cell suspension with 9.0 ml medium, and $V_{DCS} = 7.5$ ml.

   Finally, let's prepare a cell seeding suspension to seed the three flasks with the same number of cells:

   Number of flasks = 3

   Volume to be added into each flask = 8.0 ml

   Volume of cell seeding suspension $V_{css} = 3 \times 8.0 + 0.5$ ml = 24.5 ml

We will mix 7.5 ml of DCS with medium to make a total of 24.5 ml.

Finally we will distribute 8.0 ml into each T-25 flask.

2. Extraction medium is used to dissolve chemicals that leach out from test material, and the concentration of these chemicals will likely be more dilute if a larger volume of the extraction medium is used. Subsequently, the extraction test results might show milder cytotoxicity if a smaller surface area to extraction medium ratio is used. Therefore, it is important to keep this ratio as well as the extraction time consistent when comparing test results.
3. When testing a new material for cytotoxicity, we compare it to the negative and positive controls to decide 1) whether it is cytotoxic, and 2) how cytotoxic it is compared to a known cytotoxic material, i.e., the positive control. By using a moderately cytotoxic material as the positive control, we can decide if the new material is *more* or less cytotoxic than the positive control. If we use an extremely cytotoxic material as the positive control, then we would not be able to tell when the new material is *more* cytotoxic than the positive control—cells are all dead in both cases, and we end up getting less information about the cytotoxicity of the new material from the test.
4. The alloy seems to be noncytotoxic from these test results. A cytotoxicity scale of 1 from the direct contact test probably reflects the physical damage caused by the weight of the alloy slug. To confirm that cell damage observed in the direct contact test is physical rather than chemical in nature, you can perhaps use more alloy samples in the form of foils; the lightweight foils likely will cause less physical damages than the heavy slugs.
5. They should probably pay attention to: 1) The extraction test relies on accurate cell counting, so errors in cell handling can cause large errors in the results. They should first check and see if the cell seeding density was so low that the cells entered a lag phase, making it hard to control the growth of cell cultures to the same confluency later. They should also check and see whether trypsinization of the cells before counting was thorough. 2) The content of the extract of the test material can be very different depending on the extraction time, the temperature, and the amount of extraction medium. They should check and see if they used identical conditions for the extraction.
6. In Session 3, our goal is to test potential bone-replacement materials; therefore, the interaction between these materials and osteoblast cells is relevant to the biofunctionality aspect of the biocompatibility of these materials. MC3T3-E1, a preosteoblast cell line that has many characteristics similar to osteoblast cells, is suitable for the testing purpose here. The L-929 cell line, on the other hand, is a fibroblast cell line that shares more characteristics with skin cells. MC3T3-E1 is definitely the preferred cell line to use in this case.
7. Unfortunately, you cannot use a calibration curve that is established with L-929 cells for another cell line. Different cell lines have different metabolism profiles; thus, each cell line has its unique correspondence between MTT conversion and the cell number, and it is not interchangeable with other cell lines even if they are derived from the same cell type (fibroblast in this case).
8. We used an MTT assay to count cells in Session 3 on principle and out of practical considerations. On principle: Even though we are ostensibly counting cells with an MTT assay, we are actually selectively counting *proliferating* cells because only

actively proliferating cells can convert MTT to MTT formazan. Trypan blue staining does not distinguish a proliferating cell from a nonproliferating cell as long as they are both viable. Thus, the MTT assay is more suitable for cell proliferation study. Practical considerations: To count cells using a hemacytometer, the cells must be trypsinized and lifted, and complete cell suspension can be confirmed by observing the cells under the microscope to make sure that no cells are left attached. However, it is not always possible to trypsinize and lift the cells, or to observe them directly under the microscope. For example, when cells are grown inside three-dimensional scaffolds or encapsulated in matrices, it is usually not possible to count them accurately; or when cells are grown on nontransparent substrates such as titanium discs, it is also not possible to observe them under the microscope. The MTT assay is a more suitable method for these situations.

9. In the cell proliferation study in Session 3, we did not designate Teflon as the negative control, but it does serve as one. On the other hand, do notice that the function of negative control in a cell proliferation study is different from that in a cytotoxicity study. In a cytotoxicity study, Teflon is not expected to exert any cytotoxic effect, and thus as a negative control it is the "benign control"; in a cell proliferation study, Teflon is not expected to support cell growth because its hydrophobic surface prevents cell adhesion in the first place, and thus as a negative control it is the "harmful control."
10. A cell proliferation study using the MTT assay will be suitable for this project. The following is an outline of an example experiment:
    a. Seed L-929 cell cultures in 6-well plates. Make triplicate plates to obtain statistically accurate results.
    b. When the L-929 cell cultures in the 6-well plates reach near confluency, incubate the cells with the drug at different concentrations. Use serial dilution to dilute the drug from the 10 *mM* stock solution: First add 1.8 ml cell culture medium to the wells of one of the 6-well plates, and then add 0.20 ml of 10 *mM* drug solution to the first well. After thorough mixing, take 0.20 ml from the first well and add to the second well, and after mixing, take 0.20 ml from the second well and add to the third well and so on until the fifth well. The drug concentration from the first to the fifth well is 1.0 *mM*, 0.1 *mM*, 0.01 *mM*, 1 μM, and 0.1 μM, respectively. Use the sixth well as a blank control by adding 0.20 ml medium. Use the same serial dilution procedure for the other two plates. Incubate the plates with the drug for 24 hours.
    c. After incubation, remove the medium, which contains the drug. Add MTT with fresh medium to the cells and incubate for 2 hours, during which MTT is converted to MTT formazan by the cells. Afterwards, remove the medium, dissolve the MTT formazan crystals with DMSO, and measure the absorbance of the MTT formazan in all samples.
    d. For each drug concentration, calculate the average absorbance and standard error using the triplicate data. Plot the results with error bars using the drug concentration as the x axis, and the absorbance as the y axis. If necessary, an MTT calibration curve for L-929 cells can be used to convert the absorbance data to number of cells, and the data can then be plotted using the number of cells as the y axis. The plots will illustrate whether the drug is cytotoxic, and if it is, at what concentration(s) the drug becomes cytotoxic.

Module IV

1. The collagen has usually been desalted during the dialysis step in the purification process; therefore, the collagen solution contains no salt ions other than ions from the acid. The addition of HBSS serves several purposes: 1) it increases the ionic strength of the collagen solution so that its osmolarity matches that inside the cells; 2) it adds the right combination of ions for maintaining the viability of cells; 3) after the acid is neutralized, the balanced salts help to buffer the pH around 7.4, the optimal pH for cells; and 4) the phenol red in the HBSS serves as a color indicator for the titration process.
2. The white precipitation is likely polymerized collagen, and it is possibly due to NaOH being added too fast so that the heat from the acid-base neutralization reaction warms up the solution locally, allowing a portion of the collagen to polymerize. If a large portion of the collagen has polymerized during titration, then there will not be enough left to form a network during the subsequent incubation—hence, no gel formation.
3. Most likely you had overshot the titration and ended up with a basic collagen solution, which is detrimental to the fibroblasts—the cells probably had died. Phenol red turns to an orange-red color around neutral pH, and at pH >8, its color becomes purplish-pink/fuchsia/magenta. You probably missed the yellow to orange-red transition by adding too much NaOH at once and mistook the color change (probably yellow to purplish-pink) as a sign of neutralization.
4. The addition of NaOH and HCl as part of the titration will increase the salt concentration in the collagen solution. For example, addition of 100 μl 2.0 *M* NaOH to 5 ml of collagen solution adds ~40 *mM* of Na^+ ions. The salt content of the collagen solution has already been adjusted by the addition of HBSS before the filtration, which has a total salt concentration of ~150 *mM*. If this solution is titrated back and forth too many times with high concentrations of NaOH and HCl, the final salt concentration might become too high for the cells to survive.
5. The encapsulated fibroblasts will exert forces on the collagen fibers and contract the gel. Therefore, if the glass ring is not added, the collagen gel will shrink. In fact, in some cases the surface area of collagen gels can be reduced to less than 10% of the original size by encapsulated fibroblasts.
6. There are a couple of methods you can use to make those measurements, and you are already familiar with one of them—the MTT assay (see Module III). On Day 7, you should prepare calibration cultures with known numbers of cells, and then incubate both the collagen gels and the calibration cultures with MTT. Afterwards, remove the supernatant and add isopropanol with 0.04 *N* HCl (which is more appropriate than DMSO in this case) to lyse the cells in the calibration cultures and those in the collagen gels. The MTT formazan produced by cells in the collagen gels will dissolve and diffuse to the supernatant. After measuring the absorbance of the solutions from the encapsulated cells and the calibration cultures, the number of cells in collagen gels can be calculated using the calibration curve.
7. The role of ascorbic acid is to stimulate collagen synthesis by the fibroblast cells in the dermis equivalent and to promote collagen cross-linking, therefore it helps to enhance the dermis equivalent both in mass and in mechanical strength.

TGF-α helps to suppress apoptosis and stimulate proliferation of HaCaT cells in the epidermis equivalent; it also enhances the expression of a number of growth factors that are important to the differentiation of HaCaT cells, and promotes stratification of the epidermis equivalent (Maas-Szabowski et al. 2003).

8. The stiffening of the samples is caused by the cross-linking of the collagen in the samples by formaldehyde. Note that excessive fixation can make tissues brittle. Although cross-linking by formaldehyde enhances the mechanical strength of the materials in some way, such treatment is not used for tissue-engineered skin replacements because the cells that are encapsulated in the cultures are killed by formaldehyde fixation, thus defeating the purpose of engineering "live" materials that can promote wound healing or tissue growth.
9. For the slide preparation, the tissue samples are kept inside the embedding cassettes during treatments with organic solvents (alcohol and xylene). Most pen or permanent marker inks are soluble in these organic solvents, so the inked labels will be washed off. Pencil, on the other hand, is made with graphite, an inorganic substance that is not soluble in solvents, so the penciled labels will not stay.

Module V

1. Choices "d" and "e" are incorrect. Choice "d" is incorrect because the formation of HA is a gradual process. Intermediates OCP, ACP, and DCP are formed first, and DCP eventually converts to HA, but the kinetics is relatively slow and temperature dependent. Choice "e" is incorrect because when filtering HA, ethanol is added to help to dry the HA. HA is inert in ethanol.
2. The completion of a reaction is ultimately driven by the free energy of the reaction. When $Ca(OH)_2$ reacts with H_3PO_4, the free energy favors the formation of HA, even though it does not dictate the kinetics of the reaction. At high pH, the solubility of $Ca(OH)_2$ is indeed very low, but a small amount will still dissolve, and the free Ca^{2+} and OH^- ions will react with PO_4^{3-} ions to form HA. The solubility of HA is much lower than that of $Ca(OH)_2$. As Ca^{2+} and OH^- ions (and PO_4^{3-} ions, for that matter) precipitate in the form of HA, more are released by $Ca(OH)_2$ to maintain the Ca^{2+} and OH^- concentrations until all $Ca(OH)_2$ is converted to HA. (This effect is also known as Le Chatelier's principle.)
3. We can first assume that the mole fraction of HA is x; thus, the mole fraction of TCP is $(1 - x)$. Consider that each TCP molecule has 3 Ca and 2 P atoms, an HA molecule has 10 Ca and 6 P atoms, and that the Ca:P ratio of the mixture is measured to be 1.60, we have the following:

$$\text{Ca:P} = \frac{3(1-x)+10x}{2(1-x)+6x} = 1.60$$

The solution to this equation is $x = 0.333$.

The molecular weights of TCP and HA are:

$$MW_{TCP} = 310.18$$
$$MW_{HA} = 1004.62$$

Now we can use the mole fractions and the molecular weights of TCP and HA to calculate the weight percentage:

$$W_{HA}\% = \frac{MW_{HA} \times x}{MW_{TCP} \times (1-x) + MW_{HA} \times x} \times 100$$

$$= \frac{1004.62 \times 0.333}{310.18 \times (1-0.333) + 1004.62 \times 0.333} \times 100$$

$$= 61.8\%$$

4. The scan rate should be $\frac{70-10}{30} = 2$ (°/min) or higher.
5. There will be 10 diffraction peaks for the mixture. Diffractions from different compounds do not interfere with one another because each compound has its own crystal structure and the lattice spacing of the crystal structure determines the diffraction pattern. The relative intensities of the two sets of peaks will reflect the ratio between the two compounds, however.
6. The initial weight percentage of the salt will determine the porosity, and the grain sizes of the salt will determine the pore sizes in the composite.
7. We can first dissolve the polymer in a solvent, and then add salt grains to the polymer solution. The solvent should dissolve the polymer but not the salt. We can then cast the polymer solution and salt mixture onto an inert surface and let the solvent evaporate, which leaves a polymer membrane with embedded salt grains. The membrane is then immersed in water to leach out the salt. The result is a porous polymer membrane.
8. Let's first calculate the cross-section area A:

$$A = \pi\left(\frac{\text{OD}}{2}\right)^2 - \pi\left(\frac{\text{ID}}{2}\right)^2$$

$$= \frac{\pi(10.6^2 - 6.9^2)}{4}$$

$$= 50.85\ (\text{mm}^2)$$

$$= 50.85 \times 10^{-6}\ (\text{m}^2)$$

The stress σ is

$$\sigma = \frac{F}{A} = \frac{552.63\ (\text{N})}{50.86 \times 10^{-6}\ (\text{m}^2)} = 10.86 \times 10^6\ (\text{Pa}) = 10.86\ (\text{MPa})$$

Since the strain ε is 1.00%, or 0.0100, the Young's modulus at this point is

$$E = \frac{\sigma}{\varepsilon} = \frac{10.86\ (\text{MPa})}{0.0100} = 1.09 \times 10^3\ (\text{MPa}) = 1.09\ (\text{GPa})$$

Note that this Young's modulus is at the low end for bone. But chicken drumsticks come from very young chickens (most are just a couple of months old), so we can expect that these bones have yet to grow stronger.

9. *Note:* This problem is intended to encourage students to use research literature for the Discussion section in their lab reports.
   a. Cortical bone is also called dense bone; it is strong, mostly nonporous (cavities do exist for osteocyte cells and blood vessels), and is mainly responsible for the mechanical strength of bone. Cancellous bone is also known as spongy bone; it is porous and much weaker compared to cortical bone. Therefore, it is reasonable to compare dense composites to cortical bone, and porous composites to cancellous bone.
   b. Of the three HA-PLGA composites, the dense composite has a modulus of 1.1×10^3 MPa, the composite with 30–40% porosity has a modulus of 337–1459 MPa, and the one with 81–91% porosity has a modulus of 2–7.5 MPa. Higher porosity seems to correlate with lower stiffness, as one would have predicted.
   c. The composites that we have fabricated resemble the composite with 81–91% porosity and 800–1800 μm pore size. (The modulus that you have measured for your HA/FA-PLGA composites should also be close to the 2–7.5 MPa value reported in Table 7.4.)

Index

C

D

E

O

P

R

S

T

U

V

W

X

Y